I0617131

Alte Armatur und Ringkunst

The Royal Danish Library
Ms. Thott 290 2°

Hans Talhoffer

Dieter Bachmann • Paul Becker • Michael Chidester
Ariella Elema • Rebecca Garber • Dierk Hagedorn • Daniel Jaquet
Christian Henry Tobler

Edited by Michael Chidester

HEMA Bookshelf

This book is published as a companion to the facsimile of the Kongelige Bibliotek's manuscript Thott 290 2° produced by HEMA Bookshelf.

A version of "Talhoffer galore" was previously published in: Dierk Hagedorn. *Medieval Combat in Colour: Hans Talhoffer's Illustrated Manual of Swordfighting and Close-Quarter Combat from 1467*. London: Greenhill Books, 2018. ISBN 978-1784382858.

A version of "Six Weeks to Prepare for Combat: Instruction and practices from the Fight Books at the end of the Middle Ages, a note on ritualised single combats" was previously published in: *Killing and Being Killed: Bodies in Battle: Perspectives on Fighters in the Middle Ages*. Ed Jörg Rogge. Bielefeld, Germany: Transcript-Verlag, 2017. ISBN: 978-3837637830.

A version of "Tradition, Innovation, Re-enactment: Hans Talhoffer's Unusual Weapons" was previously published in *Acta Periodica Duellatorum* 7(1): 3-25, 2019. DOI: 10.2478/apd-2019-0001.

Published by HEMA Bookshelf, LLC.
411a Highland Ave #141
Somerville, MA, 02144
www.hemabookshelf.com

HEMA Bookshelf endeavors to respect the copyright in a manner consistent with its educational mission. If you believe any material has been included in this publication improperly, please contact us.

Editorial matter, selection, and preface © 2021

Codicological description © 2020 Michael Chidester; Translation © 2020 Rebecca Garber; Chapter 1 © 2020 Paul Becker; Chapter 2 © 2020 Dierk Hagedorn; Chapter 3 © 2020 Daniel Jaquet; Chapter 4 © 2020 Michael Chidester; Chapter 5 © 2020 Ariella Elema; Chapter 6 © 2020 Christian Henry Tobler

All rights reserved.

No part of this work may be reproduced, distributed, or transmitted in any form or by any means, including photocopying, recording, or other electronic or mechanical methods, without the prior written permission of HEMA Bookshelf and the individual authors, except in the case of brief quotations embodied in critical reviews and certain other non-commercial uses permitted by copyright law.

Version 3.0, 2021

ISBN 978-1-953683-20-5

Typeset in Libertinus Serif and Libertinus Sans, which are used under the Open Font License http://libertine-fonts.org/

Printed by Lightning Source.

Contents

Preface

There's something magical about holding a Medieval book in your hands. It's a tangible piece of history produced by craftsmen of another age—parchmenters or papermakers, scribes or printers, rubricators, illuminators, tanners, smiths, metalworkers, and binders, each applying their arts to contribute to something that might be soon discarded, or might last for a thousand years (sometimes more).

The content of such a book starts with what they wrote and drew in it, which tells us what the person who designed or ordered the book thought was important, but it goes far beyond that—we can learn about how it was intended to be used based on its size and shape and the materials used to create it, and we can learn how it was actually used based on the damage it suffered and the marks left by readers. Notes and doodles on the pages, pages pasted in or torn out, even the crayon scribbles of children—these tell us about the readers, and their relationship to the book.

In the Talhoffer facsimile, which you've all received by now, I tried to capture some of this magic and create something that is a bit more than just a leather-bound book. Based on the feedback I've already received, I think I succeeded for at least some of you. The facsimile is a thoroughly-modern creation, made by 21st century artisans using 21st century techniques, but it was built based on something old and worn and full of stories.

The purpose of this companion volume is to further illuminate some of those stories. In its pages, you will find a description of the manuscript itself, a full transcription and translation, and articles from some of the leading scholars in the field of historical European martial arts studies that highlight various aspects of the manuscript and its history. Paul Becker describes the life of Hans Talhoffer in unprecedented detail. Dierk Hagedorn discusses the many treatises of Talhoffer and their relationships to each other. Daniel Jaquet describes the judicial dueling culture and customs that Talhoffer existed in, and Ariella Elema discusses the background and history of the specialized dueling weapons that Talhoffer presents to readers. Christian Henry Tobler explains the strange collection of esoterica that forms the final, inverted chapter of the book. And my own contribution seeks to delve into the life of Konrad Kyeser and his treatise *Bellifortis*, which is the source for the war machines included in this book.

This should also serve as an update of the translation and analysis released by Jeffrey Hull in 2007,[*] an important work that introduced many people to the world of Talhoffer. With this book, I hope we can do the same.

Michael Chidester

[*] Jeffrey Hull was the first author invited to contribute to this book, but sadly he declined. His publication on the manuscript is titled Fight Earnestly: the Fight-Book from 1459 AD by Hans Talhoffer, *and can be found in many places around the internet.*

Acknowledgements

We would like to thank the Royal Danish Library, whose generosity in providing license-free scans made the project possible; the artisans at Grimm Book Bindery who produced the beautiful facsimile itself; and the contributing authors whose work is found in this companion volume. Most importantly, we thank the many supporters who took a leap of faith and contributed money to this project with no guarantee that it would produce anything worthwhile.

Manuscript Facsimile Project Supporters

AGEA Editora • Mike Scot Aiken • Frédéric Albrand • Scott Aldinger • Kelly Anderton • Eric Artzt • Athena School of Arms • Eric Avila • J. B. • Megha Baikadi • Jack Baker • Jayson Barrons • M. Leoncini Bartoli • Manuela Beltrami • Joe Berry • Christopher Bertell • Hildo Biersma • Thomas Biliter • Florian Binder • Gabriel Boivin • Matthew Boltz • Devon Boorman • Andrea Bowes • Alina Boyden • Tomas Brosnan • Kendra Brown • Christoph Busche • Charles Buschmann • Stano Buštor • Daniel Cadenbach • David Casserly • Charles Flynn Castellanos • Chris Castle • J. Ceirante • Ayden Chalmers • Charles Lin • Robert Charrette • Janus Christiansen • D.G. Church • Samuel Ciortea • Gordon Clayton • Louise Clayton-Hatch • Lonnie Colson • Benjamin Conan • Peter Concannon • Christy R. Conley • Conner Craig • John Criner • Hana Croke • Myles Cupp • David D'Antonio • Rachel Daugherty • Dean Davidson • Chantelle Davie • Aaron Dean • Bas Doeksen • Matthew Drozd • Dominic Eberle • Miles Ehmling • Patrick Eldridge • Clive Emerson • Scott Farrell • Philipp Fehringer • Jessica Finley • Robert Fisher • Max Fishman • Robert L. Foster • Philip L. Frank • Kevin Frost • Alexander Fürgut • Nathan Gede • Rolf Geissbühler • Joseph Giuliano • Rebecca Glass • Tyler Glenn • David Glier • Adrian Glover • Alwin Goethals • Teck Hock Goh • Bill Grandy • Michael Green • Christopher Greenway • Dion Groot • Slade Donnie Gulan • Rohalt Gysemans • Eric Haagenstad • Peter J. Haas • Johanna Hagenah • Scott Hanson • Merlin Hartley • Ethan Heilman • Hunter Heinlein • Greg Henrikson • Joseph Hillen • Aaron Himmler • Robert Hinds • Historica Clothiers • Daniel Hladik • Stephen Holdeman • Steven Hradsky • Tom Hunt • Harry Hupman • Ben Iglauer • Ben Jarashow • Kenneth Johansen • Barry Johnson • Jimmy Kalait-zidis • Alexander Kalywas • Ville Kastari • Ben Kerr • Thomas Kessler • Keysers Ghesellen • Adrian Kim • Don Kindsvatter • Jason Kingsley • Sharon Elizabeth Kitzmiller • Daniel Kline • Alexander Knapp • Hugh T. Knight • Arne Koets • Ryan Kohler • Matthys Kool • Aitor Bleda Korndorffer • Alex Kotarakos • Robert Koulakjian • Stefan Kuizenga • Drew Lackovic • José Juan Laguarda Pradas • Anthony Laird • Michael Lammer • Daniel Lancaster • Patrik Las-ota • M. L. Leung • Alexander Hereford Lewis • Craig C. Ligon • Kenneth Vander Linde • Joseph Loder • Jeremy Loose • Jose Sebastian Lopez Rodriguez • Tavian Lucas • Magnus Lundborg • Julian Maddox • Carey Martell • A. McAuley • Bennet McComish • Dionna Mc-Grail • C.K. McGraw • Ken McKenzie • Cameron McLachlan • Stephen McMillen • Greg Mele • Emanuel Meyer • Kristopher Micozzi • Jerrod Miller • Nikolas Miller • Eliot Mook • Sean Morgan • Tobias Mulzer • Connor Neagle • Keith Nelson • Julian Nickerl • Noble Science Academy • Hans Nordström • Jacob Norwood • Maxim Nossevitch • Wolfgang Ott • Christian Ouellet • J. P. • Chris Matt Pappathan • K. Parrish • Jennifer Perry • Stefan Peterson • Andrej Pfeiffer-Perkuhn • Amanda Pitcher • Jaap van Poelgeest • Martin Pulchart • Pike Pullen • Tobi Putzo • Domenik Radke • James Reilly • Nancy D. Reimers • Ashton Ricks • Alexander Ripa • Anthony Rischard • Gregory Rodriguez II • Julian Ronneberger • Eetu Röpelinen • John Rothe • Yvonne Rusch • Inmaculada Reina Santiago • Berry Schmaal • Herbert Schmidt • Marik Schoenke • Thomas Schratwieser • Morro Schreiber • Simone Schüssler • A.J. Sedlmair • Darrell Simon • Trevor Sinz • Michael Snijders • L.R. Spivack • Derek Stack • Petra Stockbroekx • S. L. Stocki • Leopold Stoffels • Brian F. Stokes • Robert Sulentic • Tim Szczesuil • Thomas Sznigir • Spencer Tai • Tim Teino • Taweechai Thongrod • Peter Törlind • Lu Torres • Aron Travis • Triangle Sword Guild • Thomas Tyndan • Charlie Underwood • Burak Urgancioglu • Thorsten Urhahn • Kevin Vartanian • Pierre Viangalli • Warhorse Studios • Roland Warzecha • Tobias Weidler • Andreas Weiss • Jesse Whitfield • Guy Windsor • Christopher Wolfla • Jeffery Wood • Jo York • Bart Zantingh-van Yperen • Conrad Yu • Jaku Zalesak • Marijn Zuidgeest • Jarred • Ross

Codicological Description[1]

Michael Chidester

Copenhagen, Royal Danish Library; 1459; Southwest Germany.

Materials and physical description

Ms. Thott. 290 fol. is a paper manuscript with 150 leaves in folio format, measuring 300 mm × 210 mm, plus one smaller leaf added during the 18th century. Strips of parchment were sewn into the center of each quire to reinforce the paper, a common practice in early paper manuscripts. Folia 16, 68, 75, and 118 were damaged, and later repaired with patches glued on-to 16ʳ, 68ᵛ, 75ʳ, and 118ᵛ (see fig. 1).

Its binding, which appears to be original, consists of wooden boards covered in blind-tooled parchment,[2] with bronze centerpieces and corner fittings with integrated bosses. All eight corner pieces are preserved, but the centerpiece from the front cover is lost; there is also evidence on both covers that it once had a single clasp (see fig. 2).

Figure 1: Paper patch on folio 118ᵛ, and conceptual reconstruction by Tracy Zoeller.

Dating and origin

Folio 150ᵛ is dated 1459 and an explicit on folio 103ᵛ agrees, adding that the book was completed during the festival of Pentecost (which fell on May 14th in that year). The manuscript's paper is watermarked with a bull's head with eyes as well as a single leaf with a flower with seven petals.[3] This paper was made in Northern Italy[4] from 1435-1457,[5] but the

---

[1] I must acknowledge that I have not had the opportunity to study the manuscript in person. This description is based on a close ex-amination of the high-resolution scans provided by the Royal Danish Library, supplemented by information in the catalogs of Hils (no. 27) and Leng (no. 38.3.4/39.4.9), who also never examined the manuscript, as well as more recent work by Sørensen (who did).

[2] Sørensen 160.

[3] Ibid.

[4] Sørensen 160.

[5] Hull 11.

Figure 2: The front cover, including the name "Thalhoffer".

language used is Swabian[6] or Swabian-Alemmanic,[7] suggesting an origin in southwestern Germany. This lines up well with the other two manuscripts by Talhoffer with known places of origin, Königseggwald[8] and Württemberg.[9]

Provenance

Hans Talhoffer's name appears on folia 10[v], 101[v], and 103[v], and his last name is also written on the front cover. It is likely that he was the original owner after its creation in 1459. It was acquired by Count Otto Thott in the 18th century, who glued an extra leaf (I) to folio 1[r] containing a table of contents. Thott bequeathed his collection to the Royal Library upon his death in 1785, where the manuscript has resided ever since.[10]

Content

The manuscript can be divided into four major sections: prose fencing teachings by Hans Talhoffer (partly modified from Johannes Liechtenauer); a heavily abridged copy of a German translation of the disordered 7-chapter version of Konrad Kyeser's *Bellifortis*; illustrated fencing teachings with minimal captions by Hans Talhoffer; and an anonymous assortment of short esoteric writings. See the table (facing) for more information.

Collation and foliation

The manuscript consists of three quires, each containing 26-27 bifolia.[11] In its present state, three leaves are missing from the beginning of the first quire, one each from inside the first

---

[6] LENG 52.

[7] HILS 74.

[8] Talhoffer, Hans. Untitled [manuscript]. Ms. XIX.17-3. Königseggwald, Germany: Königsegg-Aulendorf Collection, ca. 1448-59.

[9] Talhoffer, Hans. Untitled [manuscript]. Cod. icon. 394a. Munich, Germany: Bayerische Staatsbibliothek, ca. 1467.

[10] SØRENSEN 159.

[11] Quires of this extreme size, which are 4-6 times larger than those typically seen in this period, are characteristic of clerical records. When found in other types of manuscripts, they typically indicate that the manuscript was not created by a professional scribe, but rather by a clerk moonlighting. However, we have no other indication of Michel Rotwyler's profession to corroborate this. His surname merely indicates a family origin in the Free Imperial City of Rottweil.

1^{rv}   Anonymous poem on fencing
"Zorn ort der brust czu bort"

1^v   Prose text on fencing by Hans Talhoffer
"Item die glos der rechten kunst nach dem als die maister die abgeteilt hand"

2^r-5^v   Talhoffer's revision of Liechtenauer's recital on the long sword
"Hie lert der talhofer ain gemaine ler in dem langen Schwert von der zetel &c"

8^r-10^v   Procedural description of judicial dueling
"Hie vint man geschriben von dem kempfen"

11^r-11^v   Portraits of Talhoffer
"der schribt an ain knuo faden"

12^r-15^r   Original illustrations combining tools from *Bellifortis*

15^v-48^v   Anonymous translation of the 7-chapter *Bellifortis* by Konrad Kyeser (excerpt)
"Dyser strittwagen sol nauch diser form mit geschmid gevestnet sin"

49^r-60^v   Illustrated wrestling plays by Hans Talhoffer
"Item zulouffend ringen uß den armen"

61^r-71^r   Illustrated dagger plays by Hans Talhoffer
"Für den obern stich Mit dem linggen arme"

71^v-74^v   Illustrated unarmored poleaxe plays by Hans Talhoffer
"Den slag versetzen und hertz abstossen"

75^r-79^r   Illustrated mixed weapon plays by Hans Talhoffer
"Mit dem schwert fur den Slag mit dem spieß"

80^r-84^r   Illustrated duel between man and woman by Hans Talhoffer
"Hie schlecht er nach dem fuoß"

84^r-94^r   Illustrated armored dueling by Hans Talhoffer
"Der anfang des kampfs"

94^v-97^r   Illustrated unarmored mounted plays by Hans Talhoffer
"Daz anryten zu den vynden"

97^v-101^r   Illustrated longshield dueling by Hans Talhoffer
"Der erst anlouff mit schilt und schwert nach schwäpschen Siten"

101^v-102^r   Portrait and heraldry of Hans Talhoffer
"Hie Maister Hanns Talhofer"

102^v-103^r   Illustrated longshield dueling by Hans Talhoffer
(uncaptioned)

103^v   Explicit by Michel Rotwyler
"anno domini 1459"

104^r-110^r   Illustrations of weapons and clothing for dueling
"Der schlilt hert zu Dem Kolben"

110^v-117^r   Illustrated longshield dueling by Hans Talhoffer

117^v-123^v   Illustrated sword and buckler plays by Hans Talhoffer

124^r-130^r   Illustrated unarmored and armored mounted plays by Hans Talhoffer

130^v   Illustration of armored jousting

131^r-137^v   Illustrated armored poleaxe plays by Hans Talhoffer

138^r-139^r   Illustrated wrestling plays by Hans Talhoffer

140^v-142^v   Treatise on anatomy, possibly a German translation of *Liber nonus ad Almansorum*
"Hie fähet an ain buoch, und daz da saget wie der lyb innwendig gestalt sye."

142^v-148^v   Treatise on astrology, *Planetenkinderlehre*
"Hie stand geschriben von saturnuß"

149^v-150^v   Hebrew and mathematical diagrams and lists

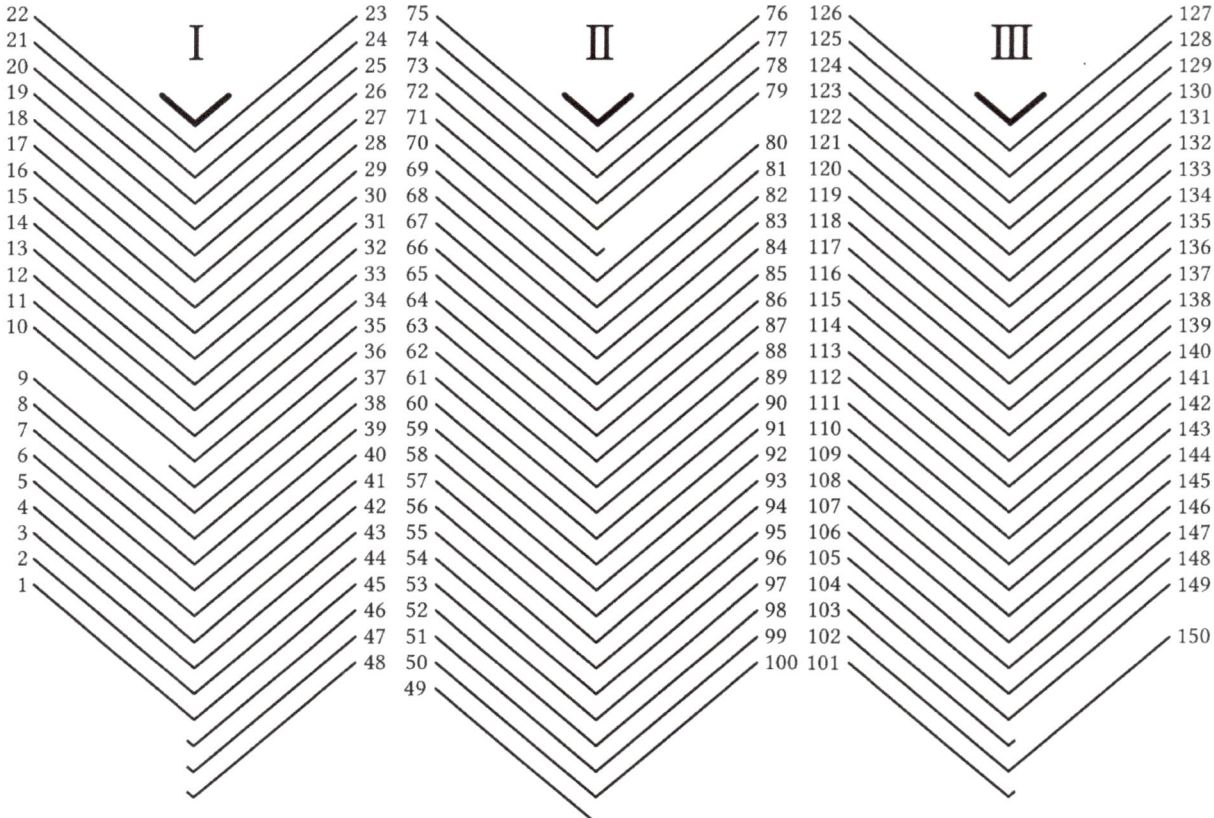

Figure 3: Collation of the manuscript.

and second quires, and one from the end of the third quire.[12] The collation is thus:

$$XXVI(-4)_{48} + XXVII(-2)_{99} + XXVI(-2)_{150}$$

The loss of these six leaves occurred before the folio numbers in the upper right corner were added in the 18th century. A diagram of the current quire structure can be found in fig. 3. The endpapers also appear to have been torn out, and a small leaf was added to the front in the 18th century (I + 150).

Unusually, the final ten folia of the manuscript are written upside-down, with their contents beginning on folio 150^v and continuing in this way until 140^v (followed by the blank 40^r). Due to the quire structure mentioned previously, it is clear that this was an intentional decision (and not, for example, a case of one quire being accidentally turned upside down during a rebinding).

Writing and decoration

The manuscript is written in bastarda by two hands. The primary hand can be identified as Michel Rotwyler's based on the explicit on 103^v. The secondary hand, found on 10^{rv}, 93^v, and 107^v-110^r, might be an autograph from Talhoffer himself

---

[12] SØRENSEN 159-160.

(based on the signature on 10ᵛ). There is also a third, much later hand which added brief notes on 79ᵛ and 80ʳ.

The single name "Thalhoffer" is written on the front cover in a contemporary script (see fig. 2), but this matches none of these three hands.

The pages generally have no text or illustration frames; ff 1r-10ᵛ, 103ᵛ, and 140ᵛ-148ᵛ constitute the text-only sections, whereas 11ʳ-103ʳ, 104ʳ-139ʳ, and 149ᵛ-150ᵛ contain some mixture of text and diagrams or illustrations. The text-only portion of the manuscript consists of 9-33 lines of text with rubrics in red ink, including three-line Lombards on 2ʳ, 9ᵛ, 141ᵛ, 142ʳ, 142ᵛ, 143ᵛ, 144ᵛ, 147ᵛ, and 148ᵛ, as well as two-line Lombards on 141ᵛ, 143ʳ, 144ʳ, and 146ʳ. The illustrated portions include 2- to 10-line captions in the *Bellifortis* material and 1- to 2-line captions in the fencing material, either above or in the center of the illustrations.

Illustrations

The illustrations are generally pen and watercolor drawings, 150-180 mm tall, possibly by Clauss Pflieger.[13] Folia 15ʳ-48ᵛ, 104ʳ-110ʳ, and 149ᵛ are drawn in portrait format, whereas 11ʳ-14ᵛ, 49ʳ-103ʳ, and 110ᵛ-139ʳ are landscapes. *Bellifortis* includes drawings

Figure 4: Illustration from folio 130ʳ.

---

[13] HULL 10.

of castles, war machines, and tools, whereas the fencing sections generally show two human figures in attitudes of fighting with each other, sometimes in a ring and sometimes in empty space. Folia 104^r-110^r show weapons and clothing for dueling, and 101^v and 149^v seem to be full-body portraits of Talhoffer and an anonymous Jewish teacher. The pen drawings are high-quality and lively, but the painting varies in quality, sometimes flat and sometimes incorporating fine shading and color gradients.

Bibliography

HILS, HANS-PETER. *Meister Johann Liechtenauers Kunst des langen Schwertes.* Frankfurt am Main/Bern/New York: Peter Lang, 1985.

HULL, JEFFREY. *Fight Earnestly: The Fight-Book from 1459 AD by Hans Talhoffer.* The Association for Renaissance Martial Arts, 2007. http://www.thearma.org/Fight-Earnestly.htm. Accessed 08 November 2020.

LENG, RAINER; HELLA FRÜHMORGEN-VOSS, NORBERT H. OTT, ULRIKE BODEMANN, PETER SCHMIDT, CHRISTINE STÖLLINGER-LÖSER. *Katalog der deutschsprachigen illustrierten Handschriften des Mittelalters. Band 4/2, Lfg. 1/2: 38. Fecht- und Ringbücher.* Munich: C. H. Beck, 2008.

SØRENSEN, CLAUS. "Et senmiddelalderligt tysk fægtemesterhånd-skrift på Det Kongelige Bibliotek. Ms. Thott 290 2°". *Fund og Forskning i Det Kongelige Biblioteks Samlinger* **50**: 159-189. doi: 10.7146/fof.v50i0.41246.

Transcription and Translation

Translation by Rebecca Garber
Transcription by Dieter Bachmann, revised by Rebecca Garber

This manuscript exists at the liminal period when Middle High German was becoming Early New High German. Some terms appear only in the MHG dictionaries, are sometimes mentioned in the ENHG dictionaries as "obsolete", or appear with a similar meaning in ENHG, while other concepts exist only in ENHG. Those concepts that appear in only one language period are easy to deal with. The problem arises with those that appear in both, but with different meanings. This is the natural result of changing literary patterns, cultures, and tastes, which directly impact the types of texts that the dictionary researchers can mine. However, it does call for a great deal of interpretation to tease out the meaning, particularly as the manuscript includes different types of texts, with highly varying levels of punctuation. This translation is thus one interpretation, based on my extensive knowledge of a language that was changing rapidly at the time these texts were composed. Others may find a different interpretation based on the same words, and I look forward to discussing their conclusions.

—

[Three missing leaves]

1^r

Zorn ort der brust zu bort	Wrath point, which drills through at the breast
zu baiden siten uber schiessen	Over-shoot at both sides
wecker wil stan	Crooked wants to stand
tribern strichen wil gan	[Who] drives forward with slashing wants to move
In der rosen im rädlin	In the roses[16], in the little wheel/circle,
zuck die treffen git guote sinn	Pull the hits offers good sense
krump how dem muol zu	Crooked strike at the mouth,
Im eyn flechten hab nit ruo	Do not pause at weaving in at him
Im krieg so machstu griffen	In the war, thus you can seize.
ochß pflug Darinn du nit wyche	Ox, Plow, do not be soft in these
Mit dem reiß ort schertz	Jest/shorten with the chasing point
Im schrack ort hab ain hertz	Have the point on the heart in the Frightening
Im ysen ort verwend	Use the point in the Iron
aim biffler tue fälen biß behend	Use Deceits very quickly at a Buffalo
Ekomen nach reissen ist der sitt	Chasing is the practice for coming before
Schnellen uber louffen und den schnit	Over run quickly and the slice
Daz ist ain gemaine lere	That is a common teaching
Daran dich kere	The wise teach you in this
Daz tund die wysen	Who are able to praise the art.
Die kunst kunden brysen	
Wiltu dich kunst fräwen	If you want to enjoy this art,
So lern die toplirten höwe	Then learn the doubled strikes
wer nach gaut slechten höwen	Whoever follows poor strikes,
Der mag sich kunst wenig fröwen	That one can enjoy the art but little
Auch so sind vier leger	Also, you should likewise consider
Die soltu mercken eben	The four guards
tuo Darin nit starck vallen	Do not fall strongly therein,
oder er laut darüber schallen	Or it will ring out loudly above [you]
wa man dir anbind wil	When someone wants to bind you an
So wind die kurtz schnid für	Therefore, wind the short edge in front.

1^v

Beschliessung der höw	**Conclusion of the strike**
Wiltu daz dirß fechten glück	If you want to be happy in your fighting
biß frisch verhalt nit lang die stück	Be lively, do not maintain[17] the plays for long,

---

[16] Possibly a compass rose?
[17] *Verhalten* has multiple meanings, including "to lurk", "wait for", and "to hold fast to, maintain something".

|Darzu hypschlich lachen
|und die ernstlichen machen
|Daz trow im Schwert
|Die der talhofer lert
|Im Schwert soltu nyemen trowen noch gelouben

|So rint dir daz blut nit uber die ougen etc.

|Jtem die Sloß der rechten kunst nach dem alß die maister Die abgetailt hand von genähe wegen daz ouch billich zu behaltent ist wen diß ist der recht grunde.

|Jtem zu dem ersten wenn du in ernst mit ainem fechten wilt, so luog wie du mit im abredst und uff wolche stund. Nach dem so richt dich nach notturft mit allem züg, und daz tuo selber haimlich und sag nymant waß du |im sinn habest oder tun wöllest, wenn die welt ist also |valsch. Und richt diu hentschuch nach deim vortail mit |allem züg, Schwert und wameß, hosen, und waß du |den bruchen wilt. Und merck aber wie du mit im |abredst, dann darby wirt eß beliben, wan daz schwert |hat sunst kain recht, dann daz es aigner und fryer will ist.

|Item wen du in den schrancken kumpst und an gan wilt, |so lauß yederman sagen und tun waß er wöll, und sich |nit hindersich, und hab den ernst im sinn. Und waß er mit |dir red, da ker dich nit an, und ficht ernstlich für dich daz, |und lauß im kain ruo, und trou und folg der kunst. Furcht nit sine |sleg, und wil er ernstlich an dich, zuck innß treffen dz ich |How wider fröhlich daran spricht |**Hanß** talhofer der guot man. |Daz muß er für die warhait jehen / |**W**ann eß ist im ouch |wol eben dick und offt beschehen ~

2[r] **Hie lert der talhofer ain gemaine ler in dem langen Schwert von der zetel etc.**
Wiltu kunst schowen
So vicht gelingg^gen recht mit howen
und lingt gen Rechten
Ist daz du starck wilt fechten
wer nach gaut slechten höwen
Der tarff sich kunst wenig fröwen
How nahnt waß du wilt
lauß kain wechsel an din schilt
zu dem koppf und zu dem lib
Die zick ruorn ouch nit vermyd
Nun ficht wyt mit gantzen liben
waß du starck wollest triben
Nun merck aber furbaß
und verstand ouch gar rechte daz

In addition, laugh in a pretty way
And be serious [about serious things]
The trust in the sword,
Which Talhoffer teaches
You should neither trust nor believe anyone in the sword
So that your blood does not run over your eyes.

Item: The palace of the correct art, according to which it is also easy to understand, as the masters divided it from nearby paths, because this is the correct foundation.

Item: Regarding the first. Whenever you seriously want to fight with someone, then observe how you arrange [a meeting] with him and at which hour. Afterward, equip yourself according to need with every thing, and do this yourself in secret. And tell no one what you have in mind, or would want to do, because the world is the false. And equip your gloves[18] according to your advantage with everything, sword and gambeson, hose, and whatever you then want to use. And consider, however, how you arrange [a meeting] with him, as it will remain thereby, because the sword has no rights otherwise, except that it is its own free will.

Item: when you enter into the barriers[19] [at the field of combat] and want to begin, then let everyone say and do whatever he would want. And do not look behind you, and remain earnest in mind, and whatever he says to you, do not engage with it. And fight seriously there for yourself, and allow him no rest, and trust and follow the art. Do not fear his blows, and if he seriously wants to approach you, then pull the strikes from him so that any strike joyfully gainsays that. Hans Talhoffer, the good man, he must affirm the truth of that, because it has also happened to him likewise frequently and often.

Talhoffer teaches here a general teaching about the long sword from the *Zettel*, etc.
If you want to behold the art,
Then fight to the left, with cuts from the right,
And to the left against the right
If you want to fight strongly.
Whoever follows after with poor cuts
May find little joy in the art.
Cut close to what you want,
Do not allow a Changer on your shield,
At your head, at the body
Do not neglect/avoid Tag hits
Now fight widely[20] with the whole body
Whatever you strongly want to do
However, additionally remember and also
Understand it correctly

---

[18] Or "gauntlets"
[19] *Schranken/Schränken* means to close off a space by blocking entry and/or exit. Lexer lists one possibility as specifically relating to tournaments [*turnieren*].
[20] With full extension

ficht nit obnen lingg so recht bist	Do not fight above left if you are right[21]
far nach zwayen ding	Before/after, two things [that]
sind aller kunst ain ursprung	Are an origin for all art
Din schwöch und din sterck	Remember your weak and your strong
Din arbait darby eben merck	Likewise your work therein
So machstu lern	Thus you can learn
Mit fechten dich erwern	To defend yourself with fighting
wer also erschricket gern	Whoever is thus easily frightened
Der sol kain fechten nymen lern	Should never learn any fighting.
Der höw sind fünff und haissent funff focal	The cuts are five and are called five vowels
Die lern recht und mercks fürwär	Learn them correctly and remember them well
und dar von komet unß der rechte grund	The correct foundation comes to us from them
Daz ist lützeln fechtern kund	Which is known to very few fighters.

2ᵛ **Die tailung der kunst nach dem text den nähsten weg zu dem mann zu schlahen oder zu stossen** — **The division of the art according to the text the nearest path to strike or to thrust at the man**

Zorn how du krump wer	You Wrath cut, Crooked weapon
how schihler mit schaitler	Cut, Squinter with Skull
aulber versetzt	Fool, Counter act
nachraissen höw letzt	Chasing cut [is] last,
überlouffen bind wol an	Overrunning, bind on well,
nit stand luog waß er kan	Do not stand, look at what he is capable of
Durchwechsel zuck	Change through, Pulling
Durchlouff hendtruck	Run through, Press[22] hands,
wind in die blössin	Wind[23] into the exposures.
Slachfach straich stich mit stössen	Hit catch slash thrust with jabs.

Das ist vo dem zornhow der underschid — **This is the distinction about the Wrath cut[24]**

wer dier Oberhowt	Whoever cuts at you from above
zorn how ortt dem trowt	Wrath cut threatens that one with the point
und wirt erß gewar	And if he becomes aware of it
Nymß obnen ab vnd folfar	Take it away above and move to the end
biß stercker wind wÿder	Be stronger, wind against it[25]
stich sicht erß // so nymß nider	Thrust, if he sees it, take it down
Daz also eben merck	Likewise note this:
Ob sin leger sy waich oder hört	Whether his guard is soft or hard.

Jn dem far nach
hört an krieg sy dir nit gauch — **Move afterward simultaneously[26]**
If the war begins, do not be quick

waß der krieg rempt	Whatever the war makes room for
Der wirt obnen nider geschempt	He who is above becomes ashamed below
Du machst in allen hewen winden	You can wind into all strikes
Im how ler stich vinden	Learn to find the thrust within the cut
ouch soltu Mit	Also you should consider with [that]
Mercken stich oder schnit	Thrust or slice
In allen treffen	In all meetings
wiltu den maister effen	If you want to ape/imitate the masters.

3ʳ **von den vir plößen** — **About the four exposures**

vier plöß wisse	Know the four exposures
Der hab acht so schlechtu gewisse	Pay attention to them so you will certainly strike
nit slach ungefar	Do not strike imprecisely
lug eben wie er gebar	Watch likewise how he acts.

---

[21] Probably "right-handed"

[22] Or "crush"

[23] Or "twirl"

[24] Or "This is the about the Wrath cut, which slices apart."

[25] Or "wind more broadly"

[26] I.e. to the action described next

Die vier plöß brechen

wiltu dich rechen
Die vier ploß künstlich brechen
obnen toplir
unden recht mutir
So sag ich ouch dier fürbaß
stand vest und biß nit laß
und erschrick ab kainen man
stand und sich in ernstlich an
hastuß denn recht vernomen
zu dem slag mag er nit kumen

Breaking the four exposures

If you want to avenge yourself
Break the four exposures with skill
Double above
Transmute below correctly[27]
I also say to you with emphasis
Stand firmly and do not be tired[28]
And jump back [in fear] from no man
Stand and observe him seriously
If you have correctly comprehended
He will not be able to come to the strike.

von krumm // wiedre schnyd da kumm

Werff. krum uff sin hende
Slach den ort nach sinr lende
und darby wol versetz
Mit schaitler vil höw letst
how uff sin fleche
so tuostu in schwechen
wenn eß knilt obnen
So nym ab Daz wil ich loben
und wer krum zu dir how
Durch wechsel du in schow
wil er Dich Irren
Der krieg in verfieret
Daz er nit nympt war
wa er ist ungefar

About the Crooked [cut]; how the slices come from there

Throw. Crooked [cut] onto his hands
Hit the point at his loin/haunch
And thereby counter act well
With Skull [cut], allow many cuts
Cut at his flat
So that you weaken him
When it clashes above
Then take away, I will praise that
And whoever Crooked cuts at you,
Change through, YOU look at him
If he wants to lead you astray
The war misleads him
So that he does not perceive
Where he is not exposed to danger.

3ᵛ Die tailung der kunst nach dem text den rechten weg und die uß richtung der zwierhin

Die zwierh benympt
waß von dem tag her kympt
und die zwierh mit der stercke
Din arbait darby mercke
zwierh zu dem pfluog
zu dem ochsen hart gefuog
waß sich wol zwircht mit springen
Dem mag ouch gar wol gelingen
Den fälar darmit fiern
unden uff mit wunsch her rieren
verkere mit zwingen
Durch louff ouch mit ringen
Den elenbog nym in der wäug
und mach den fäler nit träg
zwifachß fürbaß
schnid lingg yn und biß nit laß

The division of the art according to the text about the right way, and the instruction about the Crosswise cuts

The Crosswise takes away
Whatever comes from the Day.
And the Crosswise with the strong.
Be aware of your work with that.
Crosswise to the Plow
With great skill to the Ox
Whoever strikes crosswise well with springing
May also succeed very well
To guide the Deceit[29] with [the Crosswise]
From below upward to touch at will
Reverse with force
Running through also with wrestling
Take the elbow in the scales
And do not carry out the Deceit slowly
Continue doubled
Slice long inward and do not be slow.

Daz ist die uß richtung von der schillherin

Schylher ain bricht
waß püfler schlecht oder sticht
wer von wechselhow drowt
Schilher daruß in beroubt
Schlecht er kurtz und ist dir gran

This is the instruction about the Squinter cuts

Squinter breaks into
Whatever a buffalo strikes or thrusts
Whoever threatens from a Changeover cut
Squinter robs him of it
If he strikes short and that is of little importance to you

---

[27] Or "to the right"

[28] Or "slow"

[29] *Fehler* is the opposite of a *Treffer*, which is something that hits or succeeds. The *Fehler* is the losing throw in dice, the missed shot in archery and shooting. It is, however, an action that *might* hit, but it is assumed that it will *miss*.

Durchwechsel so gesigest im an
Schilh zu dem ortte
Nym den halß ane forchte
Schilh zu der obern
schaittel Schlach starck wil er din baitten

So machst du in wol betöwben
4ʳ Die fallerin kunst berowben

Change-through defeats him
Squint to the point
Take the throat without fear
Squint to the top
Of the skull, strike strongly if he wants to delay yours.
Thus you can deafen[30] him,
[And] steal the nonexistent art.[31]

Daz ist von dem schaittler Die ussrichtung etc.

Der schaittler dem anttlüt ist gefar
Mit siner kur der prust vast gefar
waß vom im da kumpt
Die kron daz ab nympt
Schnyd durch die kron
So brichstu sie gar schon
Die straich truck
Mit schniden ab zuck

This is the instruction about the Skull [cut], etc.

The Skull cut is dangerous to the face
With its curve, strongly dangerous to the chest
Whatever comes from it.[32]
The Crown takes away
Slice through the Crown
Then you break it already
Press the strokes
Pull away with slices.

von den vier leger

Vier lege alain
Davon halt und flüch die gemain
ochß pfluog aulber
vom tag sindz dir nit unmer

About the four guards

Hold only to the four guards
And curse[33] the common
Ox, Plow, Fool,
From the Day should not be unknown to you

von den vir versetzen

vier sind versetzen
die die leger ouch ser letzen
vor versetzen hiet dich
gschichtz dir nott eß miet dich
ob dir versetzt ist
wie daz dar komen ist
So merck waß ich dir raute
strych ab haw schnell und draute
setz an vier enden an
blyb stan und besicht den man

About the four counter actions

[There] are four counter actions
That also greatly damage the guards
Guard/Protect yourself from counter actions
If this plight happens to you, it will repay you,
If you are counter acted,
How that this has happened
Thus note what I advise you:
Slash up, strike fast and quickly
Apply [it] to the four ends
Remain standing and visibly study the man.

4ᵛ Daz ist von dem nach raisen

Nach raisen lere
verhow dich nit zusere
Sin höw recht vernymmp
Din arbaitt dar nach beginn
und brüff sin geferte
ob sie syent waich oder hörte
nun lerne in daß
den alten schnit mit macht

This is about the chasing

Learn about the Chasing
Do not strike too much, too often
Correctly identify his strike
Your work begins afterward
and check/test his movement
whether they are soft or hard
now learn in that
carry out the old slice with [that].

von dem überlouffen

wer des lybß unden remet
Den uber louff der wirt obnen nider geschemet

wenn eß plitzt oben
sterck eß daz ger ich lobn
und din arbaitt mache
oder truck eß zwifache

About the overrunning

Whoever clears away[34] the body from below
Overrun him, he will be shamed downward from above
When it sparks above
strengthen it, I desire to praise that
and do your work
or press it twofold.

---

[30] Or "destroy"
[31] Or "skill"
[32] Whatever comes from the Skull cut.
[33] Or "flee"
[34] *remen => räumen*, meaning "to make space, clear away, evacuate".

vom absetzen

kanstu die rechtu absetzen
all höw und stich sie dir letzen
der uff dich sticht
driff den ort daz im bricht
von baiden sitten
triff allemal darzu schritten

About the setting aside

If you know them correctly, use setting aside
[For] all strikes and thrusts which are impeding you
Whoever thrusts at you
Hit the point that breaks[35] it[36]
From both sides
Hit every time, step for this purpose.

vom Durch wechsel

Durchwechsel lere
von baiden siten stich nit sere
Der uff dich bindet
Durch wechsel in schier findet

About the changing through

Learn the Changing through
From both sides, do not thrust hard
The one who binds you above
Changing through finds him quickly.

vom zucken alle treffen

tuo nahet eyn binden
Die zucken gend gut finde
zuck trift erß tzuck mer
arbait er find in dut im we
5ʳ   zuck alle treffen
den maistern wiltu sie effen

About pulling all hits

Do the Pulls close to the binding on
End well to find
[The] Pull, if he hits it, pull more
If he works, find him, injure him
Pull all hits
If you want to ape the masters.

vom durch louff

Durch louff lauß hengen
Mit dem knopf griff wilt ringen
wer gegen die stercke
durch louff damit mercke

About the running through

Run through, let hang
You want to wrestle with the pommel grip
Whoever [comes] strongly against you
Remember the Running through with that.

vom abschniden

Schnid ab die hörte
von unden in baiden geferte
vier sind der schnit
zwen unden zwen obnen mit

About the slicing off

Slice off the hard ones
From below in both movements
The Slices are four
With two below and two above.

vom hennd trucken

Din schnyd ver wende
in die zwierh gar behende
und gang nach an den man
stoss mit dem ghiltz schon
wiltu denn nit schallen
so hastu zway eynfallen

About pressing[37] hands

Reverse your edge
Very quickly into the Crosswise
And move close to the man
Jab already with the hilt
If you don't want to scream[38]
Then you have two attacks.

von den zwain hengen

wer dir zestarck welle sin
heng fall im oben eyn
zwey hengen werden
uß ainer hand von der erden
Doch in allem geferte
So machstu sin waich oder hörte

About the two hangings

Whoever wants to be too strong for you
Hang, attack him from above
Two hangings arise
From one hand from the ground
Yet in all movements
You can thus be soft or hard.

von sprech venster

Sprechfenster mache
stand frolich besich sin sache
wer sich var dir zühet abe
Slah uff in Daz eß schnape

About the spreading[39] window

To conduct the spreading window
Stand joyfully, observe his actions
Whoever moves himself at you, pull away
Strike him on top so that it snaps

---

[35] Or "counters"
[36] Or "him"
[37] Or "crushing"
[38] Or "yell"
[39] Or "spraying", or possibly "speaking"

14

5ᵛ luog und schüch kain man
eß schatt nit waß er kan
hastuß recht vernomen
zu dem slag lauß in nit komen

Die besliessung der zetel
wer wol bricht
und endlich gar bericht
und brichz besunder
Jeglichß in drü winden
wer wol hengt
und winden dar mit bringt
und wint mit achte
mit rechter betrachte
Darmit ir aine
Die winden selbdritt ich maine
so sind ir zwaintzig
und vier zel sie aintzig
von baiden siten
Acht winden mit schrite
spricht hannß talhofer
got lauß unnß aller schwer

Look and shoo away no man
It does not damage whatever he can
If you have correctly understood this
He cannot come to the strike.

The conclusion of the *Zettel*
Whoever breaks well
And finally narrates/informs
And breaks it especially
Everything into three windings
Whoever hangs well
And adds windings to that
And gains with eight
With correct consideration
So that each one,
[Of] the windings themselves I consider to be triple
Thus, their numbers are twenty-
And-four singularly counted
From both sides
Eight windings with steps.
So says Hans Talhoffer,
God keep us all from perjury/bodily injury.

6ʳ [Blank]

6ᵛ [Blank]

7ʳ [Blank]

7ᵛ [Blank]

8ʳ hie vint man geschriben von dem kempfen

Item wie daz nun sy daz die decretaleß kempf verbieten, So hat doch die gewonhait herbracht von kaisern und künigen fürsten und hern noch gestatten und kempfen laussen, und darzu glichen schierm gebent, und besunder und umb ettliche sachn und artikeln, alß her nach geschriben staht.

Here one finds writing about fights.

Item: Even though it now is the case, that the decretals forbid fights, yet the custom established [in the past] by emperors and kings, princes and lords, [was] to grant and allow fights, and additionally to offer the same protection, and, as is subsequently written here, particularly about several topics and subjects.

Item zu dem ersten maul daz Im nymant gern sin Eer laut abschniden mit wortten ainem der sin genoß ist Er wolte Er hebet mit im kempfen wie wol er doch nit recht wol von im kem ob er wolte und darumb so ist kämpfen ain muotwill

Item: Regarding the first point: that no one likes to have his honor loudly cut short with words by someone who is his associate. He would rather fight with him, even though he could justifiably walk away from him if he wanted, and therefore fighting is [an act of] free will[40] [that is legally unjustified].

Item der sachen und ardickelen sind siben Darumb man noch pfligt zu kempfen

Item: There are seven topics and subjects for which one is still obligated to fight.

Item daz erst ist mortt
Daz ander verräterniß
Das dritt ketzery
Daz vierd wölher an sinem herrn trulos wirt
Daz fünfft umb fanknuß in striten oder sunsst
Daz sechst um valsch

Item: the first is murder.[41]
The second is betrayal.[42]
The third is heresy.
The fourth whoever becomes unfaithful to his lord.
The fifth embracing strife or the like.
The sixth false representations.

---

[40] *Muotwille* is the drive to carry something out, which can equally have positive or negative roots, justifications, outcomes. According to Lever and Grimm, in the case of legal writing only, *mutwille* is represented as "the opposite of that which is demanded by law".

[41] *Mort* = murder = premeditated killing in secret in ENHG, as listed in Grimm. However, there is also influence from the French *morte*, which simply means to kill (without the association with murder).

[42] Or "treachery"

Daz sibent da ainer junckfrowen oder frowen benotzogt

The seventh, whoever treats a virgin or woman violently.

Item spricht ain man den andern kempflich an, der sol komen für gericht und sol durch sinen fürsprechen sin sach für legen, darumb er in denn an clagt und sol den man nennen mit dem touff namen und zu namen. So ist recht, daz er in für gericht lad und in dry stund beklag uff dryen gerichten nach ain ander. kumt er denn nit und veranttwurt sich nach nymant von sinen wegen, so mag er sich fürbaß nit mer veranttwurten, er bewyse dann Ehafte nott als recht sy, so sel man in verurtailen alß fer in daz fürbott innerhalb landes begriffen hant. Je dar nach, alß die ansprach ist gegangen, darnach sol daz urtail ouch gan.

Item: If one man formally challenges the other to single combat,[43] he should come before the court and should submit his cause through his legal representative, as to why he then brings a suit against him, and should name the man with Christian[44] name and surname. Then it is correct, that he [the accuser] summons him [the accused] before the court and accuses him three times at three courts[45] in succession. If he does not come and no one responds on his behalf, then he himself can no longer respond from that point onward, unless he proves legitimate necessity, according to the law. Then one should announce the judgment about him, to the same extent that the legal summons has been expressed in words[46] within the country. Depending on where and when the accusation was submitted, the judgment should also be carried out there.

Item der da kempflich angesprochen wirt uff den dryen gerichten und er ainost zu der antwort kumpt und leugnot darum man in an gesprochen hat und spricht er sy des also unschuldig und der sag uff in daz nit war sy und daz wöll er widerumb mit kempfen beherten und uff in daz wysen alß denn recht sy un dem land darinn eß sy und forttert dar über mit urtail seinen lertag, so werdent im sechß wochen ertailt zu sinem lertag und vier tag von dem gericht werdent im auch ertailt, daruff sie kempfen süllent alß in dem land gewonhait und recht ist.

Item: Whoever has been formally challenged to single combat at the three courts, and comes at any time to respond, and disavows that about which the [other] one has challenged him, and he states "he is thus innocent of that", and the [accuser] says to him, "that is not true", and he wants to assert that again with combat and prove that on himself, according to the law in the country in which this occurs, and demands, in addition to the judgment, his training time.[47] Then he is granted six weeks as his training time, and four days are also granted to him by the court, on which days they are to fight as is the custom and law in the country.

Item versprechent sich zwen man willkürlich gen einander ain kampfez vor gericht, den git an auch sechs wochen lertag und sol in frid bannen baiden, und wolcher under den den frid brech, uber den richtet man on den kampf alß recht ist.

Item: If two men randomly promise to fight against one another before the court, then one also gives them six weeks of training time and both are commanded to keep the peace, and whichever among them breaks the peace, he is condemned without the combat, as is correct.

wie ainer dem anderen mit recht ᵘᵝgan mag

How one can legally withdraw from [fighting with] the other.

Item ist daz ein man kempflich angesprochen wiert von aim der nit alß guᵒt ist alß er, dem mag er mit recht uß gan ob er wil oder ob ain man echtloß gesagt würde oder worden wer, dem mag man ouch des kampfes absin.

Item: If a man is formally challenged to single combat by one, who is not as good as he [the challenged] is, he can legally withdraw from [fighting with] him if he wants, or if one man were said to be legally deprived of rights,[48] or has become legally deprived of

---

[43] *Kämplich ansprechen = die förmliche Herausforderung zum Zweikampfe* according to Lexer, which includes the *"provocatio ad certamen"* (challenge to combat) according to Grimm. This is not mere insult. This is a formal declaration.

[44] Lit. "baptismal"

[45] This is not three different locations, but three successive meetings of the same court.

[46] Or "writing"

[47] *Lehrtage* is a specific period of time prior to a single combat. Grimm: LEHRTAGE, *m. plur. lehrzeit, lernzeit: (man soll)* jhm sein lehrtag zum kampf zugeben, ... nemlich sechs wochen und drei tag. *kolbenrecht bei* Schottel 1238.

[48] I.e. has been called/labeled illegitimate, unauthentic, has lost legal rights derived from legal birth or marriage.

Item spricht aber der edler den mindern an zu kempfen, so mag der den minderen nit wol absin.

rights, one may also withdraw from the fight with him.

Item: However, if the nobler formally challenges the lesser to single combat, then the lesser cannot withdraw.

9ᵣ **Item wie aber zwen mann nit mit ainender mügent kempfen und wolcher wil under den den zweye dem andern wol uß gan mag ~**

Item: How it is possible that two men may not fight with one another, and whichever one of the two wants to, can withdraw from [fighting with] the other.

Item wenn zween mann gesinnt sind biß uff die fünffte sipp oder näher die mügent durch recht nit mit ein ander kempfen und des müssn siben man schwern die die vatter und muotter halb mäge sind. ~ ~

Item: If two men are related to one another up to the fifth degree of kinship or closer, they may not legally fight with one another. and seven men must swear to this who are blood relatives on the fathers' and mothers' sides.

Item wie aber ainer dem anderen kampfes absin mag mit solichem gelimpf alß hie geschribn stät

Item: How it is possible that one can be relieved from combat with another with appropriate conduct such as is described here.

Item ob ain lamer man oder einer der böse ougen hett und kampfes an gesprochen wirt der mag sich der auch wol behellffen und dem gesunden usgan, eß sy denn daz wyse lüt daz gelich nach der ?person machen und daz müessent wyß lüt uff ir eid tun und daz also glich machen. Es mag auch der lam oder mit den bosen ougen wol einen an ir statt gewinnen der für iro ainen kempfe. ~ ~

Item: If a lame man, or one, who has bad eyes, and is formally challenged to single combat, he can naturally acquire aid for himself with that, and withdraw from the healthy one. Unless wise people make this equal according to the persons, and wise people must do this according to their oath, and thus make them equal. It can also be [the case that] the lame man or the one with the bad eyes wants another to prevail in their place, who would fight for that one.

Item wenn also die sechß wochen uß sind und der letst tag komen ist den in der richter beschaiden haut daruff kempfen sullen, so sullen sie beide für den richter komen mit solichem ertzögen und in solich acht alß die gewonheit und das recht lert in dem lande dar inn sie kempfen sullen oder nach dem alß sie mit einander gewillkürt habent Item etc. ~

Item: When the six weeks are thus up and the last day has arrived, on which the judge has decided that they should fight, then the two should come before the judge with such a display of honor[49] and in such a spirit[50] as custom and law teaches in the country in which they are to fight, or according to which they have confirmed through free will with one another.

Item so soll da der cleger schweren daz er der sach darum er dem ainen man zugesprochen haut schuldig sy und denn so sol man in ainen ring machen und grieß wartten und urttail geben

9ᵥ nach wyser lute raut und nach des landeß gewonhait. Und wer uff den tag in den ring nit kumpt den urttailt man sigeloß in irre denn Ehafte nott die sol er bewysen alß recht ist ~ ~

Item: Then, the plaintiff should swear there, that he makes the accusation about the subject about which he challenged the other man, and then one should make a ring, and [find] judges[51] for the legal single combat, to render judgement according to the advice of wise people and the customs of the country, and whoever does not appear in the ring on that day, that one is judged [to be] without victory [and] in error, unless [there is] legitimate necessity, which he shall prove according to the law.

Hie staut wie man sich halten sol wenn die kempfer in dem ring komen sind uff die stund und uff die zit so man pheindiglich kempfen sol

Here it is written how one should hold one's self, when the fighters have entered the ring at the hour and at the time [at which] one should fight as adversaries.

---

[49] *Ertzögen*: possibilities from Lexer: 1) could be the participial form from *erziehen* [Lexer] drawing a sword specifically upward, or raising/training humans and animals. 2) *erzöugen* = to show/demonstrate. 3) *er zöugen* = *ehre zeigen* to display honor. Possibilities from Grimm: 4) *erziehen* = to raise a sword, to swing a sword or axe, to educate, bring up, train up, to lift up.
[50] Or "mood"
[51] Or "overseers"

WEnn die kempfer also in den ring komen sind So sol der richter von stund an alle stür und ler vestecklich verbieten by lyb und guot und sol nicht gestatten daz man einem für den andern nicht zulege und sel inß beiden machen so er imer gelichest mag ungenerde.

When the fighters have thus entered the ring, Then the judge should strictly forbid all disruptions and coaching from that hour onward, under threat to body and property, and should not permit that one is not assisted or punished more than the other, and should make the two of them as equally impaired[52] as he can.

Das ist was recht wer ob der kempfer ainer uss dem ring fluch oder getriben wurd

Item wolcher kempfer uss dem ring kumpt Ee denn der kampf ain ende haut Er werde daruß geschlagen von dem andern oder fluche daruß oder wie er daruß käme oder aber ob er der sache vergicht darum man in denn mit recht an gesprochen haut, den sol man sigeloß urttailen. Oder wolcher den andern erschlecht und ertötett der haut gesiget.

Dem sol man aber richten alß des landes gewonhait und recht ist darumb sie dem mit ainander gekemppffet hand.

This is what would be legal, if one of the fighters would flee or be driven out of the ring.

Item: Whichever fighter exits the ring before the combat has an end, [whether] he is struck by the other or flees therefrom (or however he would exit therefrom), or whether he admits to the issue about which [the other] one had legally summoned him, that one should be judged to be without victory. Or whichever one slays and kills the other, that one has won. However, one should judge him according to the custom and law of the country about the issue regarding which they have fought with one another.

—

[One missing leaf]

10ʳ Nun merck uff dissen puncten der ist notturfftlich zu uerstend

Item des ersten so soltu den maister wol erkennen der dich lerren wil dz sin kunst recht und gewer sy und dz er frum sy und dich nit veruntruwe und dich nit verkürtz in der lerr und wiß die gwer zü zerbraitten da mit er kempfen wil.

Now take note of these points: it is imperative to understand [them].

Item the first: You should be well acquainted with the master who wants to teach you, that his art is correct and true, and that he is honorable[53] and does not treat you disloyally and does not abbreviate[54] in teaching you, and knows to prepare the weapons with which he wants to fight.

Och sol er den maister nit uff nemen
er schwer im dann sin frumen zwerbent
und sin schaden zwendent
deß glich sol er dem maister wider um sweren
sin kunst nit witter zleren.

He should also not accept the master,
[unless] he then swears to him to strive for his honor
and to turn away from his evil.
He should in turn swear the same to the master
not to teach his art to others.

Hie merck uff den maister

Item der maister der ain understat zu leren, der sol wißen daß er den man wol erken, den er lerren wil, ob er sie schwach oder starck, und ob er gäch zornig sy oder senftmüttig, och ob er gütten auttem hab oder nit, och ob er arbaitten müg in die in die ?harr; und wenn du inn wol erkunet haust in der lerr, un wz arbait er uermag dar nach müstu in lerren dz im nütz ist gen sinen vind. Och sol der kempffer und der maister sich hütten dz sie niemand zu sehen laussend und in sunder sie gwer da mit sy arbaittent und sich baid hütten vor vil geselschafft und von dem vechten wenig sagen dz kain abmercken da von kom.

Here [is a] note to the master.

Item: The master, who has a subordinate[55] to teach, he should know that he is well acquainted with the man whom he wants to teach, whether he is weak or strong, and whether he is quickly angry or gentle, also whether he has good breathing or not, also whether he can work at length, and when you have become well acquainted with him in teaching, and [know] what work he is capable of, then afterward, you must teach him that which is useful for him against his enemy. Also, the fighter and the master should guard themselves, that they allow no one to see [them] and particularly the weapons with which they work. And the two should guard from a lot of socializing and say little about the fight, so that no discovery is made about that.

---

[52] Or "injured"

[53] *vrum* from Lexer includes pretty much all of the virtues of a knight: *tüchtig, brav, ehrbar, gut, trefflich, angesehn, vornehm, wacker, tapfer,* that is, capable, virtuous, honorable, good, excellent, respected, genteel (well-bred), brave, and courageous.

[54] Or "abridge"

[55] I.e. student.

von kuntschafft

wie der kempffer und der maister kuntschafft möchte hon zu rem widertail, wz sin wesen wer, ob er sy strarck oder swach, ob er och sy gechzornig oder nit, und wie sin touff nam hieß, ob man wölt dar uß bracticiern oder rechnen. Ws ist och nottürfftig zu wissen wz maister in lerr dz man sich dar[nach] müg richten

wenn er nun gelert ist und in den schrancken sol gon

So sol er zu dem ersten bichten, dar nach sol im ain priester ain meß lesen von unßer frowen und von sant jörgen, und der priester sol im segnen sant johanns myne und dem kempfer geben. Dar nach sol der maister in ernstlich versüchen

10[v] und inn under richten dar uff er bliben sol, und sol in uff kein ding haissen acht hon dann uff sin vind, und den ernstlich an schowen.

Merck uff dz infüeren

Item wenn der man kompt in den schrancken so sol er machen mit dem rechten fuß ain krütz und mit der hand aines an an die brust und sol fürsich gon im namen des vatters und sunes und des hailigen gaistes. Dann sind in die grieß[wartten] nemen und sind inn fürren gegen der sunnen umbhe. So sol dann der kempfer die fürsten und herren bitten und [die] umb den kraiß stand dz sy im wölle helffen got bitten dz er im sig wölle geben gegen sinem vind und alz er war und recht hab.

Dar nach sol man in setzen in den sessel

Wenn er nun gesessen ist so soll man im fürspannen ain tüch und sin bar hinder im an den schrancken und sine gwer sind wol gehenckt sin und gericht nach nottürfft

Die grieß wartten oder täpffer

Der maister und die grieß wartten söllend mercken uff den richter oder uff den, der den kampff an lauffen wirt. Wann der rüfft zu dem ersten mal, so sol er den

About scouting. [56]

How the fighter and the master can have knowledge regarding their opponent, what his character is, whether he is strong or weak, whether he is also quickly angry or not, and what his baptismal name is, whether one could predict or calculate from that. [57] It is also necessary to know, what master is teaching him, so that one can direct oneself accordingly.

When he is now taught and should enter into the barriers. [58]

Firstly, he should then confess. [59] Afterward, a priest should read him a Mass about Our Lady and Saint George, and then the priest should bless Saint John's wine [60] to him and give [it] to the fighter. Afterward, the master should seriously examine him, and instruct him as to where he should remain, [61] and [he] should pay attention to nothing except for his enemy and [should] view him seriously.

Note regarding the entry.

Item: When the man enters the barriers, then he should make a cross with the right foot and one with the hand at the breast, and should proceed in the name of the Father, and of the Son, and of the Holy Ghost. Then the judges of the single combat are to take him and are to guide or direct him around [the ring] counter to the direction of the sun. [62] Then, the fighter should ask the princes and lords and those standing around the circle that they would help him to ask God that He would grant him the victory against his enemy, as he has truth and law [on his side].

Afterward, one should set himself in the seat. [63]

When he is now seated, then one should set up [64] a cloth in front of him and his litter [65] behind him at the barriers, and his weapons are well hung and organized according to need.

The judges of the single combat or overseers.

The master and the judges of the single combat should pay attention to the executioner or to that one who will start the combat. When he calls for the first time,

---

[56] Literally "information-gathering"

[57] Probably onomancy based on his baptismal name (which is included in an earlier manuscript Talhoffer owned) and/or astrology based on his date of baptism.

[58] *Schränken* are the barriers used to form the ring/square/space for combat.

[59] This is a religious confession to a priest, in case he dies.

[60] St. John the Apostle was known for blessing a poisoned cup of wine, where the poison left in the form of a snake. *Johannesminne* or *Johanneswein* became customary throughout Germany in the 12th c. Numerous examples are found in: Hanns Bächtold-Stäubli: *Handwörterbuch des deutschen Aberglaubens*, Bd. 4, Berlin 1932, Sp. 745–760.

[61] I.e. within the boundaries of the combat ring.

[62] The sun travels clockwise.

[63] *Sessel* usually refers to a seat of honor, generally a chair with a back and arms.

[64] Or "extend"

[65] A *Bahre* is a horizontal means of carrying something, and may refer to both a *Tragbahre*, a litter for the injured, and also a *Totenbahre*, a litter for the dead. "Bier", which is a cognate, has a narrower definition than the original.

man haisen uff ston un dz tüch von im ziehen, und wann man rüfft zü dem dritten mäl so sol er in haissen hin gon und in got enpfelhen.

then he should command the man to stand up and pull the cloth away from himself, and when one calls for the third time, then he should command him to go there and commend him to God.

Von dem nach richter

Item der kempffer sel wartten das im nützit an dem lib über den ring oder schrancken uß gang dann wz dar über kem: so stat der nach richter an dem schrancken der hott imß ab mit recht ob er angerüft wirt.

About the executioner.

Item: the fighter should observe,[66] that at no time does his body move past the ring or the barriers, because whatever would extend past that, the executioner is standing at the barriers, he cuts it off with legal right, if he has been called upon [to do that].

x talhoffer

11ʳ der schribt an ain knuo faden

der schribt uß dem mund und wirt schwartz

X Talhoffer

He writes on a knotted thread/string.

He writes [the words] from his mouth and becomes black.

11ᵛ [No text]

12ʳ [No text]

12ᵛ [No text]

13ʳ [No text]

13ᵛ [No text]

14ʳ [No text]

14ᵛ [No text]

15ʳ [No text]

15ᵛ Dyser strittwagen sol nauch diser form mit geschmid gevestnet sin und mit ainem schirm alß hie gezaichnet statt und die dar uff strittent die süllent waffen haben alß hie gemalet statt.

This battle wagon should be fortified according to this shape using metal work, and with a shield, as is drawn here, and those who fight on this, they should have weapons as is painted here.

16ʳ Diß ist ain mentschlich bild gemachet von geschmid in wendig hol und sol gefült werden mit ambra und mit nägelin bluost dz sol man setzen in die örter der palaste war es sin antlüt kert da git es guoten rouch ze allin zeitten ~

This is a human image made from metal work, internally hollow, and should be filled with amber and with chervil blossoms.[67] One should set it in the places of the palace, wherever it turns its face, there will be pleasant aromas at all times.

16ᵛ Diß ist ain grosse plid mit dem man stain würffet und stett und vestin brichet

This is a large stone thrower,[68] with which one throws stones and breaks cities and strongholds.

17ʳ Diß ist ain katz und ouch ain löffel mit einem schnellen schirm mit dem man gewapnet volk in zinnen hebt. ~

This is "cat" and also a runner with a fast shield with which one lifts armored people into the battlements.

17ᵛ Diß haisset der groß rud Es gaut zu schlossen uff dem redern und sol der muren eben höhin han die gefierten bruck die dar an hanget solt mit dem sail nider lauffen bis dz das vorder ÿsen die mur begriffet dar nauch gand die sächs uß der hüh die stritten und vechtet umz dz sy behebent dz sy begeret ~

This is called the "large hunting dog". It moves to castles on the wheels and should the walls have equal heights, the square bridge which hangs thereon should lower down using the rope until the front iron grips the wall. Afterward, the Saxons go out of the hollow [interior]. They battle and fight until they capture what they desire.

---

[66] Or keep in mind.

[67] *Chaerophyllum.* Note: these are both aromatics.

[68] *Blide = Steinschleuder*, which is any weapon capable of casting stones.

18ʳ Mörck disen löffel der ist edel und guot dz obertail sol geuestnet sin mit bretter bis uff den vierden sprossen das du tribest an Die muren durch das gand gewaupnet lüt uff an die muren dz hinder tail machstu uff heben als dz vorder ~

Take note of this runner which is noble and good. The upper part should be fastened with boards up to the fourth [ladder] rung that you move to the walls. Armed people go through that up onto the walls. You can lift the rear part up like the front [part].

18ᵛ Mit diser katzen gewinet man geschwind muren mit der höhi die obnan uf gezimert ist henge der bruck mit dem sail biß dz die vorderen ysen die mur begriffent dar nach louffent die gewapnetebn in die muren ~

With this "cat", one quickly gains walls with the height, which is built at the top. Hang the bridge with the cable until the front iron grip the wall. Afterward, the armed [men] run into the walls.

19ʳ Diß ist ein hohi bruck die gaut uber wasser und uber aller graben die sol gezimert sin von hohem gezimer und sol die bruck dar an hangen dz man sy uss wendig uff hebt bis dz die reder die reder die statt begriffent so laut man die bruck vallen und mit dem val begrift dz spitzig ysen das ertrich und macht der höhi ebenhöhi machen

This is a high bridge, which moves over water and across all ditches. It should be built from a tall chamber,[69] and the bridge should hang thereon, so that one lifts it up externally until the wheels grip the location, then one lets the bridge drop, and with the dropping, the pointed iron grips the ground and makes of the height even with the height.

19ᵛ Diß haisset ain gnaper und ist ain hüpscher züg / Es gaut uff sechs redern und staut uff zwain sülen an dem zu gan so naiget es dz houpt zu der erde und richt den schwanz uff bis daß eß zu der mure kompt so richt es dz houpt uff die mur und rürt die erden mit dem schwantz und die dar under sint die schedigent die uf den muren dz ober tail von dem gnapper sol gevestnet sin mit kuder mit mist und mit grünen hüten das Inen haiß wasser für noth geschütz nit schaden müge wen mann den hinder sich züchet so züchet sich das houpt wider uff die erde und der schwanz über sich umz dz dar an die sicher statt komest ~

This is called a "rocker"[70] and is a pretty thing. It moves on six wheels and stands on two columns. At the beginning, it inclines the head to the ground and lifts the tail up until it comes to the wall. Then it raises the head up to the wall and touches the ground with the tail, and those who are under it, they injure those on the wall. The upper part of the "rocker" should be fortified with decoys,[71] with manure and with green hides, so that hot water for emergency protection cannot injure them. When one pulls it behind them, then the head pulls back onto the ground and the tail is above until you come to a safe place.

20ʳ Dise gegenwürtige höhe beschirmet daß volk in wendig hindenan und vornan und ouch obnan ~

This present elevation shields the people inside from the rear and the front and also from above.

20ᵛ Das ist ains münches kappe die im hindenan uf dem ruggen lit die furt man uf den drin redern ains vornan under dem spuz und zwain nauch gende obnan sol sy verschoppet sin mit küder und verdecket mit hüten und andren dingen die dar zu gehörend / es sol mit starckem holtzwerck zesamen gefügt sin versorget mit ysen daß es notdürftig ist das im schwer stain oder starck geschoss icht schade wan du an muren kumst so richt im den spitz uf mit der ainen schiben dem kliment hinach alle die dar inn sind ~

That is a "monk's hood", which lies behind on his back. One guides it on the three wheels, one in front under the point and two following. It should be covered above with decoys[72] and covered with hides and other things that belong with that; it should be joined using strong woodwork, supplied with iron. That is necessary so that heavy stones or hard shots do damage to it. When you arrive at the walls, then direct the point up with the one wheel. All who are inside it then climb it afterward.

21ʳ Diß ist ain brot das man haisset bis cott und ist zwürend gebachen brot. Es ist nutz uff festinen und in gezelten eß belibt gar lang zitt guot one schimel ~

This is a bread that one calls "zwieback" and is twice baked bread. It is useful in fortresses and in tents. It remains good for a long time without mold or mildew.

21ᵛ Diß hultzin laiter inwendig gemachet mit schiben sol sich zu der mur fügen mit ainem angeschlagnen sail / dar nach sich an bindet die stucke von der laiter wen du die ablegen wilt So ledige dz lincketeil von der laitern mit dem usren sail so ist sy ledig ~

This wooden ladder, made with disks on the inside, should connect itself to the wall using a cable fastened to it. Afterward, lash the pieces of the ladder when you want to remove it, then separate the left part from the ladder using the outer cable, then it is free.

---

[69] Or "space/interior"

[70] *Gnapper* = a rocker/totterer/nodding movement.

[71] Or "bait"

[72] Or "bait"

22ʳ Dise gaisel ist ain messer wen du wilt und diser näpper boret durch ain zwifaltig blecht ~

This whip is a knife when you want, and this drill[73] bores through twofold metal sheets.

22ᵛ Du solt mercken ain zu gan mit graben du solt graben krumbs umb wider und fur dar über ainen schirm setzen der sich für mit langen stangen die graben behüttent dz volk vor werffen und gand sicher zu und wenne sÿ wellent abtrëtten so söllent sÿ den schirm nauch inen ziehen so gand sÿ sicher ~

You should consider an access using ditches. You should dig crookedly around, back and forth; set a shield over it, which is guided using long poles. The ditches protect the people from [stone] throwers and [they] approach safely, and when they want to exit then they should draw the shield behind them, so that they move safely.

23ʳ Diß ist ain tötlich wer kleine stoß kerrlin gefüllet mit stainen vorannan mit scharpffen ysen die berg abge-louffen den selben mügent gewapnet lüte nauch lauffen wen die wegen wüstent mit den scharpfen ysen wz vor in ist

This is a deadly defense, little jabbing small carts filled with stones [fitted] on the front with sharp iron. Run down the mountain. Armed people can run after the same because the wagons lay waste with the sharp iron whatever is in front of them.

23ᵛ Sihe den zu gang zu den vestinen mit körben die stotzen dar under söllend in die spüz der körb gan.

See the access to the fortresses using baskets or buckets. The posts underneath should go into the point of the basket or bucket.

24ʳ wa in velsen und in hülinen volk versumet si die du sust nit über winden mügest da soltu bett oder küssi mit fedran nemen Tuo dar zu irem und pisul und zünd es an so der rouch da von gant so ersticket die in den hülinen sint also kumpst du inen zu

Wherever people are gathered in cliffs and in caves, whom you otherwise cannot overcome, then you should take mattresses or pillows with feathers; add to that urine[74] and *pisul*,[75] and ignite it so that the smoke arises from there. Then those [who] are in the caves suffocate, thus you get to them.

24ᵛ du macht machen ain liecht uff ainen turn und dar über ain laternen von rotem glaß mit ainem langen hals setz dz uff ain egk und tuo ain gros liecht dar in dz lüchtet vil milen von dir ~

You can make a light on top of a tower, and over that a lantern made from red glass with a long neck. Set that on a corner and set a large light therein. That illuminates many miles from you.

25ʳ Ufgerichte gezelt ~im~ machst du vestnen in disen weg mit höltzern obnan mit geschmid gespüzet und in die erde gesetzt mit der kluoghait über wonden die türgen den küng von ungern.

You can fortify erected tents in this way, with wooden [stakes] pointed on top with metal work and set into the ground. The Turks overcame the King of Hungary with this wisdom.[76]

25ᵛ Dise vihele sol von dem besten stahel gemachet sin mit ainem zwifalt ruggen und in wendig hol und mit blÿ gefüllet die selbe vÿle vilet so haimlich und so linse dz es nieman gehören mag ~

This arrow should be made from the best steel, with doubled ends and internally hollow and filled with lead. The same arrow flies so secretly and so quietly that no one can hear it.

26ʳ Ingenius pulcherium quo protulit equestres vide quod considera quantum quod quo quado finire

quado placeat retrahe sumtulo quo[77]

Ingenius and pleasing for horses to be swum[78]
Look at the thing, consider how many, which are this way
When it is fastened,[79] when it is pleasing, draw back with the small rope.

26ᵛ [No text]

---

[73] *Näbiger* is a broad term including both a drill and an auger.

[74] *Irem => urin*

[75] Unknown word.

[76] Or "prudence"

[77] The wrong text is included here. The original (Latin) 7-chapter version had this text on the page facing this illustration, and they were swapped by a careless scribe. The creators of Talhoffer's manuscript only included one of the two illustrations with its wrong caption, leading to much confusion. This is also the only Latin text in this version, though the Cod. 3086 (which it was copied from) has several more that weren't translated.

[78] Or "floated"

[79] Or "inserted"

27ʳ Diß ist ain zug mit dem du schwimen machst bind es für den buch des schwantzes houpt sol hangen die baine süllent lini sin und die ringgen ysinŷn ~

This is an article with which you can swim. Bind it in front of the stomach. The head of the tail should hang. The legs should be linen and the rings iron

Du ganst oder schwimest mit disem züg über ain yeglich wasser Du solt war nemen dz hopt vornan die hend und den ruggen sint verdeckt und diser züg gemachet von bukinem leder und die kand süllent sidin sin ~

You go or swim over any [body of] water with this article. You should pay attention that the head is forward on the hand, and the back is covered. And this item is made from ram skin leather and the edges should be silk.

27ᵛ Diß ist stigen mit der laiter mit ainmaligen und zwifaltigen foußstapffnn uf gan und tribt sich uf und die lang zwifaltig gabel ufenthalt die laitter undennan.

This is [a] stairway[80] with the ladder with single and double footsteps.[81] Go up and set it up, and the long double tines below support the ladder from below.

28ʳ Diß stiglaiteren sind von sailen gemacht die soltu mit diser langen gabel anschlachen

These climbing ladders are made from rope.[82] You should fasten them using the long fork tines.[83]

28ᵛ Du vergüldest ain krantz mit wissen rosen In disen weg nim goldpluomen und zerstoz die und nim zerschlagen aÿer klär und gebrenten win und mische die zestossen bloumen da mit und besch bestrich die rosen so werdent sy goldvar ~

You gild a wreath with white roses in this way. Take yellow flowers[84] and grind them, and take beaten egg white and brandy wine, and mix the ground flowers with that and coat the roses, so that they become gold colored.

29ʳ Diser schlang ist sunden wider grossen gebresten wonn er klinet uff zinnen und uf türn wenn man in uß wendig streket so gant er mit den schiben uff dar nauch sol man in und keren biß die hoken die vestnung begriffent und die spizigen ysen undennan in die erde gand ~

This "serpent" is a probe[85] against major failures, because it climbs on battlements and on towers. When one stretches it outwardly, then it moves upward using the disks. Afterward, one should reverse it until the hooks grip the fortress and the pointed iron at the bottom goes into the ground.

29ᵛ [No text]

30ʳ Diser tegen sol ain bomöl haben vier hende lang oder mer viht mit dem tegen in der rechten hand mit dem sinwelen schilt in der lingk hand.

This dagger should have a short shank[86] four hands long or more. Fight with the dagger in the right hand, with the round shield in the left hand.

30ᵛ Diser groz kolb hört zü dem schilt du solt den schilt in der lingken hand füren und den kolben in der rechten hand und mit inen baiden vehten stechen und schlachen

This large club belongs with the shield. You should guide the shield with the left hand and the club in the right hand and fight, stand, and strike with them both.

31ʳ Diß ist ain schling da mit man stain würfft und ist ain küntlich gewer mit diser wer über wand dauid golŷam und ist nutz uff vestinen und uff dem velde ~

This is a sling with which one throws stones, and is a sneaky weapon. With this weapon, David overcame Goliath, and [it] is useful on fortifications and in the fields.

31ᵛ Diß bad ist beschriben von Galieno den obrostern arzat und ist guot für manger hand gebresten besunder für dz zittren der gelider und für den gebresten das flusseß der guldin ader du solt nemen dise

This bath is described by Galieno the premier doctor and is good for many types of breakdowns, particularly for trembling of the limbs and for breakdowns in the flow of the bright artery, you should take these

---

[80] Or "steep incline"

[81] Sets of rungs?

[82] Or "cable"

[83] *Gabel* doesn't exactly mean hook, but sort of any type of forked/pointed agricultural tool.

[84] *Goldblumen* can be flowers made from gold (which appears unlikely here) or any number of yellow/gold flowers, including chrysanthemum, *calendula officinalis* (marigold), *heliotropium*, *solsequium*, areola, dandelion, *Catha palustris* (marsh marigold), *centhauria*, *aurifolium*, *asphodelus*, *heliochrysum*, *anthemis tinctoria*, *ranunculus*, *lilium martagon* (turban lily), etc.

[85] At the time, a *sonde* was a medical instrument for examining a wound or another point that could not be reached with one's hand.

[86] *Bomöl* => *baumel* = small *baum*. In the context of fighting/dueling, a *baum* is a staff, so a short staff.

krütterbugg weremuot. baldrion bertram Eindorn benedict // heid agrimonien harwe und gewinne die bÿ schönem wetter wesche es wol und schnid sÿ ze stucken und leg die in dem hafen under dem bad dz der tampff durch dz ror in gange wenn man denn dar inne geschwitzet so sol man die krütter uß dem hafen nemen und sich da mit riben und weschen so mann iemer haissest mag Es ist all monat nütze on in den hundtagen

herbs: artemisia, wormwood,[87] valerian,[88] chrysanthemum or tanacetum,[89] thorny restharrow,[90] St. Benedict's herb,[91] kraut,[92] heather or broom, agrimony, and European water lily.[93] // Obtain them in nice weather, wash them well and cut them into pieces and lay them in the crock under the bath so that the steam goes in through the tube. When one has sweated in there, then one should take the herbs out of the crock and rub and wash him or herself with them, the hottest that one can endure. This is useful all months except during the dog days.

32ʳ Diß ist ain anders schling und ist nütz zeweld uff vestinen und uf bergen ~

This is another sling and is useful in the fields, on fortifications, and on mountains.

32ᵛ Diß ist ain kuchi mit ainem kemi da der rouch in allen winden schnellerlich enweg flühet nim claur von eÿern und tuo es von den tottern und güß dz clär durch ain trachter in ain blater dar nauch klopff die tüttern wol durch ain andern über sind dz clar in der beschlosse blatter dar nauch güß die tuttern dar in und süd es hert werde dz ist denn ain groß eÿn dar ab vil essen mügent ~

This is a kitchen with a chimney so that the smoke flows away quickly in all winds. Take the white from eggs and separate if from the yolk, and pour the white through a funnel into a large flat bowl. Afterward, beat the yolks well through and over one another. If the white is in the closed flat bowl, afterward pour the yolks into that. Simmer until it becomes hard. That is then a large egg from which many can eat.

33ʳ wiltu verborgen tragen brieff berlin gold silber oder edel gestain nim leÿn und mach ain holding dar uß und leg ez dar in dar nauch nim kalch und saltz und tempern dz mit aÿerclar überzüch es da mit wenn das erdrucknet so wirt es herter der ain stain ouch machtu ain brot en zwai tailen und hülen und dar in legen und wider zesemen machen oder ain hülzin holtz klotz boren und ain zapffen dar in drengen und absegen dz der zapff und der klotz ain holtzen sie

If you want to secretly carry letters, pearls, gold, silver, or precious stones, take linen and make a hollow item from that, and place it in there. Afterward, take lime and salt and temper that with egg white, [and] coat it with that. When it is dry, then it becomes harder than a stone. Or separate a bread [loaf] into two parts and hollow it out, and place it there in and put it back together. Or drill a wooden block and force a peg into that and cut off the peg (and the block is a wooden [block]).

Ain krutt haisset lolin ettlich nenient ez ninilol dz wachset gern an den stetten da mann kol ~~brenett~~ gebrent hantt wer der wurtzen ~~fuindet~~ südet in win bis der win das vierdtail in gesüdet und denn dz mit anderm win müschet dz bringt den aller sterckisten schlauff den man yemer gemachen mag und wer dz also ain [maus] moss bereit in ain ganz süder wins tut und es sich da mit verainet und vermist so wirt es als ain über treffenlicher starck schloff getranck wer dez trincket der kumet von schlauffen wegen von aller kraft und wer. und man möcht da mit nider legen ain groß volk und wiß dz diß ain haimlich sach ist

An herb is called darnel[94] (some call it *ninilol*). It grows well at the locations where one has burned charcoal. Whoever simmers the roots in wine until the wine is one-fourth of the simmering, and then mixes that with other wine, that brings the strongest sleep that one can ever have, and whoever then puts moss in a well simmering wine, and combines and mixes it with that, then it becomes a surpassingly strong sleep draught. Whoever drinks that, that one loses all strength and defenses because of sleep. And one can lay a large population low with that, and know that this is a secret thing.

33ᵛ Diß ist ain haspel gemachet von zwifaltigen redern und also geordnet dz ain tail sich dem ~~sich~~ sail zu dem andern schlüsset so du zu der ainen sitan zügest so gant ez über sich zu der andren siten wider under sich

This is a winch made from doubled wheels and arranged such that one part connects the rope to the other, so that if you pull on the one side, then it goes above itself, to the other side [and] back under itself.

---

[87] *Absinthium.*
[88] *Baldrian = valeriana.*
[89] *Bërhtram = pyrethrum.*
[90] *Eindorn = ononis spinosa.*
[91] *Benedict = herba benedicta.*
[92] *Heide = erica.*
[93] *Harwe/harwurz = nenuphar.*
[94] *Loliol = lolium*: darnel, poison darnel, darnel ryegrass, cockle, or false wheat.

34ʳ Diß ist ain louffende brugge die sich fur schübet mit den vsern sailen und mit dem mitel sail gant sÿ wider hinder sich und ist ain klügen züg und sail die sail undenan und innan also angebund[en]

This is moving bridge, that is pushed forward using the outer ropes, and it goes back behind again using the middle rope, and it is a clever article, and [one] should thus attach the ropes underneath and inside.

34ᵛ Diser büchsen schirm machet sich von holtz werck und gant uff und nider man schüsset den stain her us so ez uf gant und wenn es nider gant so mag den dehain geschoss schaden der hinder dem schirm ist ~

This box shield is made from woodworking and moves up and down. One shoots stones out of it when it moves up, and when it moves down, then no shot can injure the one who is behind the shield.

35ʳ Diß schiff loufet schnelleklich wider wasser also geschwind als ain pfärt mit ~sehl~ schnellem louff Es sol zwaÿ reder haben alz hie geformirt statt ~

This ship runs[95] fast against water, as quickly as a horse with a fast gait. It should have two wheels as are pictured here.

35ᵛ Merck wo die sunn dem gold oder liechten glantzen harnasch nach gant so sol der schilt vor gan Also über windet ain manlicher vechter sinen veÿent mit der sunnen hilff der sunnen glantz in dem gold oder in dem liechten schöne | harnasch sender gemist dar us in der veÿent ougen ~

Pay attention: wherever the sun follows the gold or brightly shining armor, the shield should move in front of that. Thus, a manly fighter overcomes his enemy with the sun's help. He sends the beams of the sun (in the gold or in the brightly beautiful armor) in a measured way from there into the enemy's eyes.

36ʳ Diß gewapnet houpt schnidet mit den oren zu ÿetwedrer siten es gant uff zwain redern und mit der zungen und dem horn sticht es und töttet und ist vechther in stritten Es sol in wendig hültzin sin und uswendig starck mit ysen beschlagen dz es nieman mit hamern noch mit axen zerhowen müge der künig porrus fürte disen schirm und leite vil siner veuÿent damit nider ~

This head, equipped with weapons, cuts with the ears to either side. It moves on two wheels and it thrusts and kills with the tongue and the horn, and is a fighter in battles. It should be internally wooden and externally strongly reinforced[96] with iron, so that no one can hack it into little pieces with hammers or with axes. King Porrus guided this shield and laid many of his enemies low with it.

36ᵛ Ain wiser stritter sol sin wegen in starcken stritt also ordnen deß ersten ain wagen nauch dem anderen der nauch zwen nebent ain ander dar nauch dreÿ dar nauch vier ye meye me naüch der linien untz dz du sÿ alle erfollest mach dez heres craft dar in tail dz roß volk und ouch dz füßfolk also tailest du allen spitz diß ordnung bruch so du ziehest in die frömde

A wise fighter should organize his wagons like this when moving quickly. Firstly: one wagon after the other, afterward two next to one another, afterward three, afterward four, increasing according to the lines until you fill them all out. Divide the army's force therein, that is, the warhorses, people, and also the people on foot. Divide all the points thusly. Use this organization when you move into foreign lands.

37ʳ Diser sporn zwingt sechs pfert und sind die formen daran halb ysin und halb stechlin ez ist zebruchen in der frömde und nement man ez ain sicherhait won ez zwinget den man vest ze siuden ~

This spur subdues[97] six horses and the shapes on it are half iron and half steel. It is to be used in foreign lands and it is taken as a means of security, because it forces the men strongly to the sides.

37ᵛ man sol machen zwen wägen uf dise form da enmitten uff gewäpnet lute strittent die wägen die vechtent die buchsen die letzent und die beschlagenen tromen Ist dz du zu wasser kumest so setze die zwen wegen nach ain ander so stand wasser halb sicher / je der wagen sol sechs reder nauch siner grossi haben und ander halb wagen lang sin die pfärt sond näch den zwain ersten redern uf zwo eln gon so sint si sicher und legent alle ding nider die stangen die da under inen sint die behüttent sy vor vallen und die bretter die da zu den siten hangent beschirment die ungewaupneten pfärd Es sol ain pfart nauch dem andern gon und zu

One should make two wagons. On top of this frame there in the center, armed people fight; the wagons, they battle; the guns, they injure, and they beat drums. If you have arrived at water, then set the two wagons following one another in series, thus you are in a safe place due to the water. Each wagon should have six wheels according to its size and should be one-and-a-half wagon lengths. The horses should follow two ells behind the two first wheels so that they are safe, and they lay everything low. The rods, which are inside underneath, protect them from falling and the boards which are hung at the sides shield the un-armored horses. One horse should follow the other,

---

[95] Or "moves"
[96] Or "covered"
[97] Or "overpowers".

yetwedrer siten ist ~~ain~~ an zwain ^{en} gnuog und sind mit zwifalter ketten zesamen geheft sin

and at each side one second [horse] is enough, [they] and are held together with twofold chains.

38^r Diser züg haisset ain krebs und ist geschmidet von ysen und gant hinder sich für sich man fürt ez uf vier schnellen redern Es schnidet vornan und hindan Es sind zwifalt sicheln von langen ysen zu yetwedrer siten an den achsen der schiben sin Es hant vier ougen wenn man die anzündet so schüsset ez stain da vor geschrottene stahelstuck als ~~ev~~ ain hagel da mit man die veÿent niderleit den züg mag man machen groß oder clain als mann denne will.

This item is called a "crab" and is smithed from iron, and moves backward and forward. One guides it on four fast wheels. It cuts in front and behind. There should be doubled sickles made of long iron at each side on the axles of the disks. It has four eyes: when one ignites them, then it shoots stones in front, crushed steel pieces like a hail, so that one lays the enemy low. One can make the item as large or small as one wants.

38^v Diser strittkarr der schrott gewäpnet lüt schenkel ungewaupnet volk vellet es mit rüren Es hant hindenan braite und vornan schmale ysen scharpff gefilet als ain sichel und man fürt es uff zwain redern ~

This battle wagon, which hacks up armored people's legs, it fells unarmored people with contact. It has iron [parts], broad in the rear, and narrow in the front, filed sharp as a sickle, and one guides it on two wheels.

39^r Diser schirm gehört zu bichsen und ist gezimert von holtzwerck und sol vornan zwaÿ claine reder haben und hinden zwaÿ grossen und dz ez dester lichlicher über berg gang es sol haben zwaÿ kurze ÿsen ainer eln lang da hindan und vornan diß ist maz an bergen und ze stigen und ouch wider ab ze ziechen dz ÿsen sol man alwegen mit langen sailen hinder sich ziechen ~

This shield belongs to the guns, and is built from woodwork, and should have two small wheels in front and two large behind. And so that it can more easily go over mountain paths, it should have two short iron [pieces] one ell long behind and in front. This is measured to mountains and for climbing, and also to pull it back again. One should always pull the iron with long ropes behind it.

39^v Diß ist ain grine lange nater vornan mit starcken hangken die sol man legen zu ainr wand an muren Es mugent bruchen die lüte Sturmen und ritter die in den muren mugent in inhin zichen wz sy begriffent wenn ~~d~~ si die sail kreffteclich ziehent ~

This is a frightful, long "viper" with strong hooks in front, which one should place at a face on [exterior] walls. The people attacking [the walls] can use it and knights, who can pull into the interior in the walls, whatever they grip, when they strongly pull on the rope.

40^r Dise bruggk dienet in zwen weg Si füret sich über land uff vier redern und schwimet in dem wasser tuo als ich vor geseit hab dise brugg ist ain gütter wagenn und vest und gelücklich uff dem wasser

This bridge serves in two ways. It travels over land on four wheels and floats in the water. Do this as I said before, this bridge is a good wagon and water-tight and comfortable on the water.

40^v Diß ist ain ander stighangk der fürt sich uff vier redern biß dz der oberhangk die mur begrift und stiget man sicher da mit türn und muren ~

This is another climbing hook. It moves on four wheels until the upper hook grips the wall and one safely climbs towers and walls with it.

41^r Also fürt wasser von ainer sitten obnan ab und zu der andern sitten wider uff Enmitten inne sol sin ain stube die dz wasser ufenthaltett ~

Thus, water guides itself from above on one side and up again to the other side. [There] should be a chamber inside in the middle which detains the water.

41^v Wilde pfärd sol man laden mit dürrem holtz dar in schwebel und bech und hartz si der sattel so geschmiere sin mit ayne claur zünd dz an trib die pfärt under die vÿent so bissent und schlachent und brennent die vigent

One should load wild horses with dry wood, including sulphur and pitch and resin.[98] If the saddle is daubed or smeared with egg white, then ignite that. Drive the horse among the enemies, thus [they] bite and strike and burn the enemies.

42^r Ein recht fürpfil sol fornan sinwel sin und dar in hol dz sol man füllen mit pulver und ain secklin dar über ziehent dz ouch mit pulver gefüllet sy zin dz hindan an und schüß in bald so brint dz für und letzt den schaft wenig und wo der schaft beheftet da schadet er gröslich / ain ander für pfil nim pulver schwebel und

A correct fire arrow should be round in front and hollow therein, which one should fill with powder, and pull a small sack over that, which is also filled with powder. Ignite it behind and shoot it quickly so that the fire burns and damages the shaft only a little. And where the shaft sticks in, it damages a great deal.

---

[98] Or "gum"

26

werch temperen mit öl binds obnan alz ain spinel zünd ez an und schüß da mit

Mit ainem grossen näper bor ain loch in ainem bom und mit ainem clainen näpper ain klain loch uf die ander siten und fülle dz groz loch mit pulver und verschlah ez vast wol mit ainem clotz zünd es zu dem clainen loch an und flüh bald da von dz zerspringt von ain ander und machet ain groz tumultus

42ᵛ Diser züg ist gemacht zu drin buchsen und schüsset ẏe aine nauch der andern ~

43ʳ nim des aller besten pulvers und leg es an diß kugeln zünd sẏ an und wirff die da von kumet grosser schad wann dz für zerbrächt sẏ und brenet gar hart ~

43ᵛ Diser züg höret under dz wasser dz huopt und der lip sond verdecket sin mit leder und wol verneit, und die ougen von glesern dar in gemachet, und mit hartz und mit bech wol versichert obnan vor dem mund sol sin ain badschwam dar under zwen, dar us du den autem vahest und wider us laussest also machtu gan und sehen under dem wasser.

44ʳ Zeglicher wiß ist diser züg, uß genomen dz dz hopt verdecket ist mit ainem schweren helm und ouch die ougen dar under vermachet als vor. Ist dz das wasser starck rinnet So soltu dich beschweren mit gewicht oder ain sail binden an ainem bom oder an ainen stok, das du dar an herus mugest komen.

44ᵛ Loliol ist genant libol samen in win gesotten und der win in andern gemischet wem du den ze trincken gist der entschlaffet und entwachet in achtagen nit. nim us den rugen grautt der kreps die fige, mische es wa mit du wilt dz machet gröslich schlaffen. nim baldrion leg den übernacht in win, wem du den zetrinken gist der entschlauffet bis man in wecket. nim den somen von muratitru, wem du dz in trank gibest der begriffet den andern bi dem har. Remedia für die vogeschribne stuck ist starcker essich in die naslöcher gegossen. welher über land ritten wil oder zu ainem wachter genomen wirt und in der schlauff beschwärt, der neme körner von strepicon und lüwe es in den mund so hört der schloff uff ~

Another fire arrow: take powder, sulphur, and work the correct mixture with oil. Bind it on top like a spindle, ignite it, and shoot with it.

Using a large drill,[99] bore a hole into a tree and a small hole on the other side using a small drill,[100] and fill the large hole with powder and close it completely with a chunk of wood. Ignite it at the small hole and flee quickly from there. It bursts apart and makes a large tumult.

This item is made for three guns and shoots one after the other.

Take the very best powder and place it in these spheres. Ignite them and throw them. Great damage arises from this because the fire explodes them and burns very strongly.

This item belongs under the water. The head and the body are covered with leather and sewn well, and the eyes are made with glass therein and secured well with resin and with pitch. A sponge[101] or two should be above in front of the mouth, from which you can catch your breath and you can let [your breath] back into it, thus you can move and see under the water.

This article is [made] the same way, with the exception that the head is covered by a heavy helmet, and the eyes under that are made as before. If the water flows strongly, then you should weight yourself with weight or tie a rope to a tree or to a pole,[102] so that you can come out using it.

Darnel[103] is called *libol* [when] cooked together in wine, and the wine is mixed into another. Whomever you give this to drink, he falls asleep and does not wake for eight days. Take the fig-leaf-shaped part from the backbone of the crab,[104] mix it with whatever you want. That makes for great sleeping. Take valerian, soak it in wine overnight. Whomever you give it to drink, he sleeps until someone wakes him. Take the seeds from *muratiten*.[105] Whomever you give this to in a drink, he grabs the other by the hair.

The remedy for the previously described articles is to pour strong vinegar into the nostrils. Whoever wants to ride across the country or to be taken as a watchman, and wants to make sleep difficult, he should take seeds of *strepiton*[106] and places them in the mouth, then the sleep quickly ends.

---

[99] See note 73.
[100] Ibid.
[101] Presumably some type of natural sponge.
[102] Or "stump", "pole", "column"
[103] See note 94.
[104] Likely crayfish.
[105] Unknown word.
[106] Unknown word.

45^r [No text]

45^v nim kupffer schlag und mach dar us ain kugel in wendig hol darnach nim ungelöschten kalch ain tail galbani ain halb tail müsch dz galbanum mit dem kalch dar nauch nim schnegken gallen in glichem gewicht und leg dz galbanum dar in dar nach nim Cantarides als du wild schnid in die höpter und die flügel ab und stoß die mit glichem gewicht kecksilbers tuo es in ain kolben vergrab es in ainem mist vierzig tag und ender dz in dem mist ye an dem .v. tag so wirt es als gold der nim die kugel und bestrich si mit dem ersten stuck und las es truck nen und wenn es trucken wirt so bestrich si mit dem anderen ding zünd si an so erlöschet si nit wilt du si aber löschen so nim ain stuck von ainem küsling waich das in essich dreÿ tag und versenck und versenck die kugel mit Im	Take copper, beat [it], and make a sphere from it (internally hollow). Afterward, take unslaked lime, one part, [and] *galbanum*,[107] one half part. Mix the *galbanum* and the lime. Afterward, take turtle[108] gallbladder in the same weight and place the *galbanum* therein. Afterward, take soldier beetles[109] (as [many] as you want), cut the heads and wings off, and push with the same weight of mercury. Put it on a club, bury it in manure for forty days, and change it in the manure every 5th day. Then it becomes like gold. Take the ball and coat it with the first part and let it dry and when it is dry, then coat it with the other thing. Ignite it, and it does not extinguish. If you want to extinguish it, however, then take a piece of a flint, soften it in vinegar for three days, and immerse the ball within it.
46^r Ich haiß philomenus und bin gemachet von er oder von kupffer Ich gib kain hitz so ich ler bin wenn ich aber gefüllet wird mit Terebinte oder mit gebrantem win und man min lip zu dem für tuot dz ich haiß wird so wirf ich für in gneist vor da mit man ainer ieglich kerzen an zünden mag ~	I am called Philomenus and am made from ore or from copper. I do not give [off] any heat if I am empty. However, if I am filled with turpentine[110] or with brandy wine, and someone places my body toward the fire so that I become hot, then I throw fire in sparks in front and one can light any candles thereby.
46^v Dis sint schne raiff die sol man in dise form machen und stro dar uff decken und ob den füssen zesamen stricken zeglicher wis machstu dz machen mit langem stro under den füssen zesamen gebunden ~	These are snowshoes, one should make them in this shape and cover them with straw and tie them together above the foot. In the same way that you do that, tie them together under the feet with long straw.
47^r Ains fürsten palast machtu wermen mit disen wol schmöckenden stucken nim vil tigel die für enhaltidt mache von dürrem holtz und kolen ain für dar under tuo dar in ambram mustg Saffran gamphor mirren alteß olibanum mastick wirouch und zwaÿerlaÿ sandel Tuo userlesen wirouch dar zu oder andre wolschmeckende ding Item leg negelÿ uber nacht in win und mornens so zerklub sÿ alle Tuo die nägeli dick in ainen tigel und güß des wines ein wenig dar an. wenne denn dise stuck haiß werdent so wirt es wol riechen.	You can make a prince's palace warm with this well-flavored piece. Take many crucibles that contain fire, make a fire under them from dry wood and coals, [and] place therein ambrosia, partially-fermented wine, saffron, camphor, myrrh, old frankincense, mastic, incense, and two kinds of sandalwood. Add select incense to that, or other well-flavored things. Item: soak cloves[111] overnight in wine, and in the morning, cut them up finely. Place the cloves thickly in one crucible and pour a little of the wine on that. When this piece becomes hot, then it will smell good.
47^v Diser gesamneten gezüg ist guot in raisen für die herren ain heftseg geisfüss ain näber ain dietrich ain pfriem ainen schnit ysen ain scharschach die stuck sind notdürftig.	These assembled items are good in riding[112] for the army: a saw with a handle, crowbar, drill,[113] hook spanner,[114] pointed instrument for drilling mounted on a handle, a threading[115] die, [and] a cutting[116] blade.

---

[107] *Galbanum* is an aromatic gum resin, the product of *ferual gummosa*.

[108] *Schneggen* could an older version of turtle, or specifically garter snakes in the north; it also means snail, but snails don't have gallbladders.

[109] *Cantharidae.*

[110] Literally *terebinth*, the turpentine tree.

[111] *Nelken* can be cloves, carnations (*dianthus*), or the pistils of flowers like lilies.

[112] Or "traveling"

[113] See note 73.

[114] Or "wrench"

[115] Or "screwing"

[116] Or "shearing"

48^r	[No text]	
48^v	[No text]	
49^r	Item zulouffens ringen uß den armen	Item: Approach to wrestling from the arms.
49^v	Item daz arm brechen und uber den schenckel werffen	Item: The[117] arm-break and [the] throw over the leg.[118]
50^r	Item das burn vassen	Item: The peasant's hold.
50^v	Item daz uß zucken und werffen vor der elenbogen	Item: The pull-up and throw in front of the elbow.
51^r	Item das durchgan	Item: The pass through.
51^v	Item der bruch uber daz durchgan	Item: The counter against the pass through.
52^r	Item daz vassen im wammeß	Item: The hold by the gambeson.[119]
52^v	Der bruch über daz vassen im wammeß	The counter against the hold by the gambeson.
53^r	Daz stuck uber den arm und inn fuoß	The play over the arm and in at the foot.
53^v	Daz werffen uber die huffte	The throw over the hip.
54^r	Der armbruch	The arm-break.
54^v	Das hinder tretten	The step behind.
55^r	Der armbruch uber die achsel	The arm break over the shoulder.
55^v	Duch am hinder tretten	Push down at the step behind.
56^r	Das achsel brechen	The shoulder-break.[120]
56^v	Das halß würgen	The neck[121] throttle.[122]
57^r	Daz versuchen durchgan oder hinder tretten	The attempt to pass through or step behind.
57^v	Der buobnwurff überß houpt ~	The servant's[123] throw over the head etc.
58^r	Stuck und bruch	Play and counter.
58^v	Daz genick vassen	The neck-hold.
59^r	Der verkertt wurff	The reversed throw.
59^v	vom man zu komen	To get away from the man.
60^r	Daz beslossen vassen	The securing[124] hold.
60^v	hinder sich der bruch für daz heben	[Grab] behind oneself, the counter for the lifting.
61^r	Für den obern stich Mit dem linggen arme	For the thrust from above, using the left arm.

---

[117] General note for verbs: verbal noun phrases (das + verb), will be translated as "the [action]", not as gerunds.

[118] *Schenckel* can specifically indicate the thigh/upper leg (its precise meaning), or simply the leg.

[119] This is specifically the piece of defensive clothing that covers the buttocks. The picture shows a short, sleeveless piece of clothing, like a jerkin, while the word refers to the longer piece of clothing, the gambeson.

[120] This could potentially be a dislocation of the shoulder.

[121] Hals refers to the soft parts of the neck.

[122] Or "strangle".

[123] *Bube* is both a male child, a servant, squire, an undisciplined man, or a professional dice player. Multiple references to *Buben* and dice (*Würfel*) in the literature make this a likely pun. Lexer; *Buobe, der würfel machet buoben vil* Ls. 3. 231,15. 480,116; Mart. 56, 91. 73,1. 206,85. Pass. *K.* 161,41. Marlg. 222, 296. 362,75.

[124] Or "encompassing"

61ᵛ	Der arm bruch	The arm-break.
62ʳ	Daz verkertt werffen	The reversed throw.
62ᵛ	Den man zu werffen mit dem tegen	To throw the man using the dagger.
63ʳ	Den undern stich weren und den arm brechen	To defend against the thrust from below and to break the arm.
63ᵛ	Den man für zu werffen	To throw the man forward.
64ʳ	Den tegen ainem Nymen mit sim tegen	To take the dagger from someone using his dagger.
64ᵛ	Der ober schilt	The upper shield.
65ʳ	Der wurff uber die hufft	The throw over the hip.
65ᵛ	Den man werffen mit gewalte	To throw the man using force/violence.
66ʳ	Für werffen und durch schiessen zuck an dich	To throw forward and to shoot through; pull towards yourself.
66ᵛ	Das fahen	The catch.
67ʳ	Der bruch darüber	The counter against that.
67ᵛ	Der wurff ubern ruggen	The throw over the back.
68ʳ	Der bruch und dot stich	The counter and killing thrust.
68ᵛ	Der under bruch und hertz stich	The lower counter and thrust to the heart.
69ʳ	Der wabet stich und bruch dar für	The bobbing[125] thrust and the counter for that.
69ᵛ	Stuck ist volbracht	Play is completed.
70ʳ	Der under schilt	The lower shield.
70ᵛ	Der mortt stich	The violent killing[126] thrust.
71ʳ	Das end stuck	The final play.
71ᵛ	Den slag versetzen und hertz abstossen	Counteract the blow and thrust away [at the] heart.
72ʳ	Den stich versetzen und in er stechen	Counteract the thrust and thrust at him mortally.[127]
72ᵛ	Den bruch und wurff	The counter and throw.
73ʳ	Daz hallß ryssen	The neck[128] yank.[129]

---

[125] Or "swaying", "rocking"; *Waben* is related to honeycomb and bees, and to the back-and-forth, up-and-down movement of water. Even the relationship to *wëben* is through bees and honeycomb, not weaving, which has different roots to get to *weben*.

[126] MHG *mort/mord* was another form of "dead", loaned/borrowed from French *morte*. However, in ENHG, it does have a more violent aspect than *tot*. It is not until modern German that *Mord* picks up the sense of "murder".

[127] *Erstechen* implies a mortal wound, beyond the mere thrust of *stechen*.

[128] Hals here indicates the back of the neck/shoulder.

[129] Or "tear"

73^v	Der nott stand fur den slag	The combative stance[130] in front of the blow.[131]
74^r	Stuck verbracht	Play completed.
74^v	Daz gwer fachen	Catch the weapon.
75^r	Mit dem schwert fur den Slag mit dem spieß	Using the sword in front of the strike using the spear.
75^v	Der gewäbet Stich	The woven thrust.
76^r	Der oberhow fur den stich	The cut from above in front of the thrust.[132]
76^v	Daz end stuck mit dem schwert für die hellen barten	The final play using the sword in front of the halberd.
77^r	Daz blenden ab dem huopt Darby der wurff *in leib*[133]	The blinding at the head, thereby the throw *into the body.*
77^v	Daz endstuck mit dem messer für die hellen bartten	The final play using the Messer in front of the halberd.
78^r	Daz versetzen gen dem spiess	The counteraction against the spear.
78^v	Ain billgerin für ain langen spieß mit sinem stab	The pilgrim using his staff in front of a long spear.
79^r	Der nott stand im messer	The combative stance in Messer.
79^v	Daz stuck *da die Hand verlohren.*[134]	The play *where the hand is lost.*
—		[One missing leaf]
80^r	Hie schlecht er nach dem fuoß *und sy trifft dz houpt.*[135]	Here, he strikes at the foot *and she hits the head.*
80^v	Hie hat er den schlag ir entwert und dem arm gefangen	Here, he has disarmed her strike and caught the arm.
81^r	Der griff nach dem halß	The grab at the neck.
81^v	hie nickt siu den man	Here, she presses the man down.
82^r	Daz halß brechen	The neck-break.
82^v	Hie macht er ain end stuck	Here, he makes a final play.
83^r	Hie wil siu in töben und er siu sÿ vellen	Here, she wants to subdue[136] him and he sees her fall.
83^v	Hie tribt daz wib ain endsstuck	Here, the woman carries out a final play.
84^r	Hie macht er end	Here, he makes an end.
84^v	Der anfang des kampfs	The start of the fight.
85^r	Hie sitzend sy bed	Here, they both are sitting.
85^v	Der stand für den schutz	The stance before the spear throw.

---

[130] The earlier meaning of *nôt* is to be threatened, hemmed in, imperiled; however, it can also be used as a replacement for the term battle/combat, as that state already implies peril and threat. *Nothstand* is a later compound, and is both a predicament/peril/threat, and a state of being under predicament/peril/threat. Interestingly, it is also used in the north in a very narrow meaning for the horizontal crosspiece that stands behind a dyke gate (*Balkensiel* as opposed to a culvert) that prevents the gate from opening.

[131] Or "The combative stance prior to the blow"

[132] Or "The cut from above forward of the thrust"

[133] Italicized text added by a later hand.

[134] Italicized text added by a later hand.

[135] Italicized text added by a later hand.

[136] *Toben/täuben:* to force, tame, subdue, deafen, with an aspect of irrational fury.

86ʳ	Hie ist der schutz versetzt	Here, the spear throw is counter acted.
86ᵛ	Der ander schutz versetzt	The second spear throw is counter acted.
87ʳ	Daz anlouffen nach dem schutz	The approach after the spear throw.
87ᵛ	Der morttschlag ist versetzt	The killing blow is counter acted.
88ʳ	Daz brendschürn	The fire scissors.[137]
88ᵛ	Daz schwert nymen	Taking the sword.
89ʳ	uß dem morttschlag den gurgelstoß	From the killing blow [to] the throat jab.
89ᵛ	Das lemen	The laming.
90ʳ	Der ruggen wurff	The throw onto the back.
90ᵛ	Die versatzung zer ryssen und daz antlut stossen	Tear the counteraction up/down and jab [at] the face.
91ʳ	Der hallß Schlag	The neck-strike.
91ᵛ	ain ander stuck	Another play.
92ʳ	Stuck und bruch	Play and counter.
92ᵛ	Daz under werffen uß den brend schürn	The throw down out of the fire scissors.
93ʳ	und daz selbig end am letsten	And finally, the same end.
93ᵛ	Hie dancket er got *und hat groß noth*[138]	Here, he thanks god *and has a great need.*
	Da lyt er tod	He lies there dead.
94ʳ	Daz tragent in die fryheit hin weg inß grab / Daz got alle gelöbig selen hab amen	The freedoms carry him away into the grave, Which god has for all worthy souls, amen.[139]
94ᵛ	Daz anryten zu den vynden	The approach to the enemies.
95ʳ	Daz versetzen im schärmitzlen	The counteraction in the skirmish.
95ᵛ	Daz schiessen an der flucht	Shooting at the rout.[140]
96ʳ	Daz armbrost und swert bruchen nach dem schutz	Using the cross bow and sword after shooting.
96ᵛ	Daz glen bruchen halb und gantz	Using the lance, half and whole.
97ʳ	Der stich uß versatzunge	The thrust from the counteractions.
97ᵛ	Der erst anlouff mit schilt und schwert nach schwäpschen Siten	The first approach with shield and sword according to Swabian customs.
98ʳ	Daz suchen hinderm schilt	Seek behind the shield.
98ᵛ	Der drit und gurgel stich	The kick and throat-thrust.
99ʳ	Daz hinder treten und hertz stich volbringen	Step behind and complete the heart-thrust.
99ᵛ	Hie ist der kampf uff dem kolben gericht	Here, the fight is carried out using clubs.

---

[137] Or "The fire (spark) protection". *Schiure, schûren: schützen, beshützen*, in connection with *schirmen; schüren:* to stimulate, poke the fire to burn higher; *scheuern, fegen* = scouring away, sweep.
[138] Italicized text added by a later hand.
[139] This is *Kittelvers*: there are 4 major beats in each line, with a lot of unstressed syllables, particularly in the first line.
[140] Or "the flight/fleeing ones"

100ʳ	Der anlauß	The initial point.
—		[One missing leaf]
100ᵛ	Daz zwierhen im Schilt	The crosswise cut into the shield.
101ʳ	und daz ynbinden	And the binding inward.
101ᵛ	Hie Maister Hanns ·;· Talhofer ·;· ~ ·;·[141]	This [is] Master Hans Talhoffer.
102ʳ	bedenck dich Recht[142]	Consider correctly.
102ᵛ	[No text]	
103ʳ	[No text]	

103ᵛ **anno domini 1459**
Item das buoch ist Maister Hansen Talhoferß und der ist selber gestanden mit sinem lybe biß daz man daz buoch nach im gemalt hat und daz ist gemalet worden uff pfingsten in dem Jar nach der gepurt unsers lieben Herrn Christi Tusent vierhundert und darnach in dem Nün und fünfftzigosten Jar schrib mich Michel Rotwyler für wär

anno domini 1459
Item. The book is Master Hans Talhoffer's, and he has proven it with his body until one painted the book according to him, and that was painted on Pentecost in the year after the birth of our dear Lord Christ one-thousand-four-hundred and after that in the fifty-nineth year. Michel Rotwylar wrote me truthfully.

104ʳ	Der schlilt hert zu Dem Kolben ~	The shield prevails over the club.
104ᵛ	zum kolben	[The shield prevails] over the club.
105ʳ	zu dem schwert	[The shield prevails] over the sword.
105ᵛ	zum schwert	[The shield prevails] over the sword.
106ʳ	zum schwert	[The shield prevails] over the sword.
106ᵛ	lern kolben	Learn the club.
107ʳ	daz gwand zu dem schilt und zu dem Kampf talhofferß an tuon	The garments to be worn [used] for the shield and for Talhoffer's fight.
107ᵛ	zu dem harnasch	[To be used] for armor.
	och gewapnet	Also armed/armored.
	zum langen schilt	[To be used] for the long shield.
108ʳ	**zu dem schilt**	**[To be used] for the shield.**
	die zwen swert hörnt zu dem kampf gewapnet	The two swords belong to armored combat.
108ᵛ	die dri degen zu dem kampff	The three daggers [to be used] for combat.
109ʳ	Die gwer bruch war zu man wil	The weapon breaks whatever one wants.
109ᵛ	Die agsten zu dem kampff	The axes [to be used] for combat.
110ʳ	dis ist ain agst zu dem kampff legt man zu stucken	This is an axe [to be used] for combat, if one lays out [the individual] parts.
	die agst zu dem kampff	The axe [to be used] for combat.
110ᵛ	[No text]	

---

[141] In a banner.
[142] In a banner.

111^r [No text]

111^v [No text]

112^r [No text]

112^v [No text]

113^r [No text]

113^v [No text]

114^r [No text]

114^v [No text]

115^r [No text]

115^v [No text]

116^r [No text]

116^v [No text]

117^r [No text]

117^v [No text]

118^r [No text]

118^v [No text]

119^r der recht not stand gen zwainen   The correct combative stance against two [opponents].

119^v [No text]

120^r [No text]

120^v [No text]

121^r [No text]

121^v [No text]

122^r [No text]

122^v [No text]

123^r [No text]

123^v [No text]

124^r [No text]

124^v [No text]

125^r [No text]

125^v [No text]

126^r [No text]

126^v [No text]

127^r [No text]

127^v [No text]

128r [No text]

128v [No text]

129r [No text]

129v [No text]

130r [No text]

130v [No text]

131r [No text]

131v [No text]

132r [No text]

132v [No text]

133r [No text]

133v [No text]

134r [No text]

134v [No text]

135r [No text]

135v [No text]

136r [No text]

136v [No text]

137r [No text]

137v [No text]

138r [No text]

138v [No text]

139r [No text]

139v [Blank]

From here, the transcription and translation will continue in reverse order (beginning with 150v) to capture the sequence of the inverted pages.

— [One missing leaf]

150v 1 2 3 4 5̃ 5̃ 6 7 8 9 // 10 11 12 13 14 15 16 17 18 19 20 21 22 23 24 25 26 27 28 29 30 31 32 33 34 35 36 37 38 39 40 41 42 43 44 45 46 47 48 49 50 51 52 53 54 55 56 57 58 59 60 61 62 63 64 65 66 67 68 69 70 71 72 73 74 75 76 77 78 79 80 81 82 83 84 85 86 87 88 89 90 91 92 93 94 95 96 97 98 99 // 100 2000 190 3000 400 500 600 700 800 900 1000 2000 3000 4000 5000 6000 7000 8000 9000 10000 20000 30000 40000 50000 60000 70000 80000 90000 100000 100000 200000 300000 400000 500000 600000 700000 800000 900000 1000000 // 1111111 // 1234567 // 1459 Jar macht mich michel rotwyler für wär

Michel Rotwyler truly made me in the year 1459.

	guldin	schilling	haller	öxttlin[143]
100000				
10000				
1000				•
100			•••	••••
10		•••	••••••	••••
1	•	•••		

150ʳ

t	sch	r	k	tz	f	p	e	s	n	
ת	ש	ר	ק	צ	ף	פ	ע	ס	נ ן	

m	l	chi	i	t	h	s	uo	h	d	g	b	a
מם	ל	כך	י	ט	ח	ז	ו	ה	ד	ג	ב	א

a	b	d	e	f	g	h	i	k	l	m	m
|	|	|	|	|	|	|	|	|	|	|	|
א	ב	ד	ע	ף	ג	ה	י	כ	ל	ם	מ

n	n	p	r	s	t	v	tz	tz
|	|	|	|	|	|	|	|	|
ן	נ	פ	ר	ז	ט	ו	ץ	צ

ch	ch	sch	ז	ס	ת	ח	פ	פ	ש
|	|	|	|	|	|	|	|	|	|
ך	כ	ש	s	s	t	h	f	ff	ss

[Nonsensical Hebrew characters][144]

Alleff	א
bed	ב
gimel	ג
daled	ד
he	ה
uaf	ו
sain	ז
hess	ח
teß	ט
iuss	י
ocumichaff	כ
slechtechaf	ך
lamed	ל
ofemem	מ
pschlossemem	ם
ocuminun	נ
slechtinun	ן
samech	ס
ein	ע

---

[143] This is obviously Arabic numerals and connected to an understanding of the decimal system. The table represents money with values 1 gulden, 33 schillings, 360 Heller, and 1440 "Öxttlin". A Gulden (or gold florin) actually corresponded to 360 Heller (a coin minted in Schwäbisch-Halle), but 30 Schillings (depending on the currency even 40 or 24, but not 33). An "Öxttlin" is apparently a quarter-farthing.

[144] This may in fact be German words written with Hebrew letters from right to left. However, I can't figure out the code and the text on the page is not German.

pe	פ
ve	ף
ocumtzadick	צ
slechtitzadik	ץ
kuff	ק
ress	ר
schin	ש
taf	ת

[Nonsensical Hebrew characters][145]

Item die alle wegen nyenen denn am lesten ann wortten זונסך

Item: They always name the very last at the very front.

—

[One missing leaf]

149[v] אי[146]

איטם	Item
הנש	hans
טלהפר	talhofer
קלש	clauss
פפליגר	pflieger
מיכיל	michel
ווטוילר	rotwyler

Hie lert der Jud Ebreesch

The Jew teaches the Hebrew language here.

149[r] Blank

148[v] **Hie stand geschriben von saturnuß, der da ist alt und kalt, unrain, hässig und nydig: also sind mine kind die under mir geboren werdent. ~ ~**
Saturnuß ist der obrost planet, und der aller untugent haftenst, und der gröst, und ist kalt und drucken, und haißt darumb saturnuß zu ein gelichnuß, als die römer alle göt by in hetten und sie anbetettend und inen ouch ir oppfer gabent und brächtentz jeglichnem in sinen thempel, der deme in siner are gemachet waß. und die romer hiessent den selben iren got und nannten in Saturnus, daz ist alß vil gespochen alß der höhste. und wenn die romer also sprachent so sass er in dem hohsten tron des himels also dz er sass über ander gött und in allen herr zu gebieten, und darumb so nanten sie in Saturnus als ainen obrost gott. und wenn sie in woltent ettwerumb bitten, so detten sie gar grosse bett an in, und detten daz ein gantz iar, und sprachent daz er uber alle gött erhoher weer, und darumb sollte man in so vil zyteß zimlichen bitten, denn der obrost gott wolten mer zyteß und langer gebettn werden, dan die andnen götte allesampt, und daz waß sin ubermuot, den er in im selber hette, darumb, daz er erhöher waß uber ander götte, denn wenn in die romer enstlichn anruoffent, so wurden sie von im erhöret, und darumb so solt er mer hoffhart geniessen, wan wir lassent von denn römern, wenn sie in an ruoffent von ettlichen sachen wegen. Daz sie

Here it is written about Saturn, which is old and cold, impure, ill-tempered, and greedy. "Thus are my children, who are born under me."
Saturn is the highest planet, and the least virtuous and the largest, and is cold and dry, and is therefore named Saturn as an allegory. When the Romans would have all gods with him, and prayed to them and also gave them their sacrifices, and would bring those into his temple, which was then made in his honor. And the Romans called the same their god and named him Saturn, which means much like "the highest". And when the Romans said that, so that he sat in the highest throne of the heaven, thus that he would sit above other gods and he would be offered in all things. And, therefore, they called him Saturn as a highest god. And if they wanted to ask for anything from him, then they would offer very great prayers to him and would do this for an entire year. And they said that he would be elevated above all gods, and for this reason one should appropriately pray to him so very much time, because the highest god wanted more time and to be prayed to for longer than all the other gods together. And that was his hubris that he had in himself about this, that he was elevated above other gods, because when the Romans earnestly called upon him, then they were heard by him, and for this reason he wanted to enjoy his pride. When we read about the Romans,

---

[145] See note 144

[146] This table is presented horizontally, but here it has been rotated 90 degrees to fit into this page layout.

dar nach kum in fünff jaren Erhortt wurdent ettwen in dryssig jaren oder ettwen mynd wanenun der selbe got der under den abgötten der obrost waß, alß die römer sprächent. und der trägost also ist auch der planete genant nach im von einer gelichnuß, wenn under den siben planeten so ist saturnus

148ʳ Der obrost und der höchste in den himeln, und doch der trägost an sinem louff. und derumb so habent ettliche menschen ain tail ir Complexion von Im und sind sanguini und flegmatici, und die sind aineß hohen muotes und sahent vil sachen an und erkünden, noch mügen keiner sache ußtrag geben, und wenn sich satturnus zu denen vermischet, die da sind sangwiny, daz machstu darby erkennen, so machet dem sangwiniß ain langß antlüt, und dem flegmaticuß ain sinvels anttlüt. Es ist ouch zu wissn alß vorgesait ist, daz sich dieselben lüte vil sachen under widen und doch keiner usstrag gebent, und sind och hohtrahent und ubermütig lüte und bedunckett sie daz in nyemant gelich sy und vermügent doch nutz für sich selber, und sind arm an zytlichem guote, und die sangwini sind mit ainem langen bartte oder antlüt und tund nit gern gotzdinst, aber die flegmatici sind göttlich an in selber und furdern gern gotzdinst, und darumb so vahent die maister kain ding an zubuwen an den stunden so satturnus rengnirt, dann sie vermamen eß werd kumerlichen volbracht, dann ob eß in ainer andern zit angehaben werde. Satturnus und sine kind sind gewonlichw rouber und morder und wenn er rengnirt, so ist gut reden mit edlen lütten und der planet ist unßer vynd in alle wege der nature und stant gen Orient und ist ain planet bosser lüte und untugenhaffter, die schwartz und mager sind und türre und ist ain planete, der mannen die nit berte habent und wyß hare, und die ire claider unsuber tragent die kind.

Die under satturno geboren werden, sind mit ainer schmale brust und trurig und hörend geren von bosen dingen sagen, und tragen gelich alß unsubere claider 147ᵛ alß schöne und vermügen sich sich nit wol mit frowen, und hat doch von nattur alle bösse ding an im Satturnus erfüllet sinen louff in dreyssig Jaren und 400 und 40 tagen und 6 stunden, und von siner höhe wegen so mag man in selten sehen. und dz sind sine zaichen: der stainbock, der wasserman, die sind kalt und trucken an irer nattur, und glichent sich dem melancholicoß an siner nattur.

Hie sagt Juppitter von siner nattur und von sinen kind, wie die geren kinder schriben und lesen und ander künst.

Juppitter ist der ander planet und der ist glückhafftig und tugenthaft warm und frisch und ist ettwen vil träg an sinem louff und höret den zu die tugent hafftig sind und ist herre der mannen die da dick bärte hand und werdent nit kal und wenn er also rengniert so gaut es frowen wol sie mit knaben gand und ist guot vor

when they called upon him because of anything, that it would scarcely be heard in five years, perhaps in thirty years or perhaps never. Because now the same god, who was the highest among the idols, as the Romans said, and the laziest. Therefore, the planet is also named after him as an allegory, because among the seven planets, Saturn is the uppermost and the highest in the heavens, and yet the laziest in its course. and because of this, some people have part of their temperament[147] from him and are Sanguine and Phlegmatic, and they are of high temperament and observe many things and learn. Yet they do not want to express or provide anything, and if Saturn is mixed with those who are Sanguine, you can recognize that by the fact that [Saturn] gives a long face to the Sanguine and a sensible face to the Phlegmatic. It is also known, as previously said, that the same people possess many things, and yet express or provide nothing, and are also proud and arrogant people, and they consider no one to be their equal, and accrue useful things to themselves, yet are poor in temporal goods. And the Sanguine are with a long beard or face, and do not happily attend Mass. However, the Phlegmatics are godly in themselves and happily promote Mass. And therefore, the masters do not begin to build anything during the times when Saturn reigns, because they warn, it will be poorly completed, compared to if it were erected in another time. Saturn and his children are usually robbers and murderers, and when he reigns then it is good to speak with noble people, and the planet is our enemy in all ways of nature and stands towards the east, and is a planet of evil people and the unvirtuous, who are black and gaunt and reckless, and is a planet of men who have no beards and white hair, and who wear their clothes untidily. The children who are born under Saturn have a narrow chest and are depressed, and like to listen to talk about evil things, and wear untidy clothes equally to nice ones, and do not behave well with women, and still have, from nature, all evil things in themselves. Saturn completes his course in thirty years and 440 days and 6 hours, and because of his high course, one can seldom see it. And these are his signs: Capricorn [and] Aquarius, which are cold and dry in their nature and are similar to the Melancholy in his nature.

Here Jupiter speaks about his nature and about his children, how the children like to read and write and other arts.

Jupiter is the second planet, and it is happy and virtuous, warm and fresh, and is somewhat lazy in its course, and listens to those who are virtuous, and is lord of the men who have thick beards and do not become bald. And when it thus reigns, women succeed who go with young men,[148] and it is good to seek

---

147 Simultaneously "disposition" and "character".

148 In this time period, *Knabe* means young man, squire, not yet a knight, and not a young boy.

fursten recht suochen der planet ist genant Juppitter zu ain gelichnuß alß die maister sprechent wie ain abgot were den die römer und ander lüt hieltent für ainen got der sie vast gewerte und in ouch vast an ruofftent für ainen hellffer und berantter und versöner wenn so die römer irem got nit sin opfer gabant und er zornig ward uber sie so so baten sie den selben got Juppitter mit grossem ernste und andacht und brachtend im ochsin opfer daz er in wider hüllffe daz sie zu den genaden wider kämen gan dem got der uber sie erzürnt waß und die wyle der göte ainer zornig waß so torsten

147ʳ sie kainen got anbeten denn den got Juppitter und Juppitter ist so vil gesprochen nach römer siten alß am hellffer wann alß die römer sprehen so halff er sinem sune So er besas den obrostten tron in irem himelrich und darum so rufften in die römer an alß einen hellffer und darumb so hat Juppitter sin Complexion und sinen loff mit denen die da haissend Sangwini und so eß den wol gat nach irem willen so hellffent si den iren och vast und ander lüten und mügent sich auch wol vast arbaitten durch der lütten willen und ain mitlyden mit in haben aber von ubermuot alß si an in selber hand so mügent sie nyemant nutzit getuon man bite sie dann ernstlichen und tue in so vil liebß hin wider Juppitter halt auch sinem louff mit denen die da sind und haissend Colerice dann die hellffent ouch den luten und den Iren und tund im doch nit gelich und tund ir ding haimlichen und sie mügent ire sachen wol heimlichen und verborgenlichen tragen und sind ouch vast getrew fründ Daz kind dz under dem planetten geboren wirt daz wirt mässig Eere und recht hat eß lieb und hat gern hüpsche claider und waß da wolschmeckt und rein ist daz hat eß gern un der hant Es wirt och barmhertzig und frölich und Juppitter hat die zaichen der sonnen den schützen und den vische Juppitter erfüllet och sinen loff in fünff jaren und fünff tagen

Hie sagt Mars von siner artt, die er an Im hat mit sinem siten, stryttig und hässig und wissent dennocht nit warumb, oder gen wiem eß ist.

146ᵛ **Mars ist** der dritt planet und ist haiss und trucke ungluckhaftig und böß und dennocht milt und mässig in sinem louff und ain planet zornig lütte und die geren kriegent und roübent und kal sind und hand krauß haar und des wenig und under dem planeten ist guot in stritt gan stele rouben und brennen und die lut wunden vond ist darumb marß genant von den wysen maister zu ainem gelichnusse alß marß von den unglöbigen waß genent ain gotte deß strytteß und wann die römer wolltent streitten so rufften sie marß an und brachten im opffer in sinem tempel und fuortten in och mit inen in daz velde da sie dann stritten wolten und alß die maister sprechent darumb so

princely law. The planet is called Jupiter as an allegory, because the masters say how he was an idol, which the Romans and other people held to be a god, who strongly defended them and whom they also strongly called upon as a helper/assistant, and advisor, and a conciliator. Because if the Romans did not offer sacrifices to their god, and [that god] became angry about this, then they would offer them to the same god Jupiter with great seriousness and reverence, and also brought him his sacrifice, so that he would help them again so that they would return to grace with the god who was angry with them, and while the god was angry, then they did not dare to pray to any got except that god Jupiter. And Jupiter is addressed so much according to Roman customs as a helper, because, as the Romans say, he helped his sons. Thus, he occupied the highest throne in their heavenly kingdom and for that reason, the Romans called upon him as a helper, and because of that, Jupiter has his temperament and his course with those who are called Sanguine, and thus it goes well, according to their desire, so that they also strongly help their people and other people, and can also strongly work through the desire of the people and have sympathy for them. However, due to the overconfidence which they have in themselves, they can at no time do otherwise for anyone, if one earnestly asks them, and they offer so much love in return. Jupiter also maintains his course with those who are and are called Choleric, because they also help the people and theirs, and yet do not treat them the same. And they do their things secretly and they like to keep their things quite secret and hidden, and are also quite loyal friends. The child that is born under the planet, that child will have moderate honor and privilege. [The child] loves, and likes pretty clothes, and [the child] likes to have that which tastes good and is pure in their hand. [The child] is also compassionate and joyful, and Jupiter has the signs of the Sun, Sagittarius, and Pisces. Jupiter completes its course in five years and five days.

Here Mars speaks about his type, which he has close to him with his bellicose and hateful manners, and yet they do not know why, or even how it is like this.

Mars is the third planet, and is hot and dry, unlucky and angry,[149] and yet generous and measured in its course. And [it is] a planet of angry people, and they like to fight and steal, and are bald and have curly hair (and little of that). And among the planets, [it] is good in fights against stellae. [They] rob and burn and injure people, and [it] is therefore named Mars by the wise masters as an allegory, because Mars was named a god of battles by the unbelievers, and whenever the Romans wanted to fight, then they called upon Mars and brought him sacrifices in his temple and also carried him with them in the field where they then wanted to fight. And when the masters speak about it,

---

[149] Or "evil"

haisset er marß dann so er under den siben planetten rengnirt so müß derselben jarß vil strit und krieg sin und wann nun marß in des sonnen gang gaut so mag man in selten sehen wenn er aber rangniret so sprechent die maister so man in sehe ob dem sonne So bedüte eß grosse niderlegung under dem adel alß daz fursten und herrn ouch ritter und knecht desselben jarß nit sollent kriegen dann sie lägen danider aber die buren hand guot kriegen denn alle ding gand vast nach irem willen und darumb die kinde die dann empfangen werden so marß rengnirt die werdent vast strittig lüte und hand die natur nit mit den die da haissent Sangwiny dann die sind gar stritbar und verherend doch dick und vil an irem stritten wenn man in aber sicht under dem sonne so so hat er etliche natur it denen die da haissent melancholici die sind still schwigent stritter und gelingt

146ʳ In wol an irem kriegen und des Jarß so er rengnirt So rengnirt gewonlich ain steren der da haisset Cometa und welich lande er gesehen wirt in dem selben lande wird on zwyfel grosse turung und und hunger dann man mag in nit in allen land gesehen wann er ist nider an den himelen und nach by der mone also daz der monen schatten in umb git Darumb man in wol gesehen mag denn so der sonne ist in dem zaichin daz da haisset Cantz oder löwo und wolcheß jar er rengnirt so ist der sonn und die mon des jarß gern bresthaftig Wölcher under dem planeten geboren wirt der wirt ?ont mit ettlicher vinsternuß alß die an der sonnen brun werden und wirt ouch untugenthafftig kriegisch und machet geren unfrid under den lüten und hat under den xii zaichen den wider und den Scorppion und ir Complexion und natur und marß erfüllet sinen louff in funffhundert und dryssig tagen

Jch Sonne sag euch in kurtzer frist, daz min schin über alle planetten ist. min uff gang gyt des tages schin, und min under gange zouget die sterne fein. und macho den menschen schön und wolgemuot, daz sonst kain ander planete duot.
DEr Sonne ist der vird planett und ist haiss und trucken und ist lusteklich ain eynfliessendes liecht allen dem daz da lept uff erde Er ist ain planet Schön und lesteklich und erluchtet den menschen sin

145ᵛ anttlüt und ouch den luten mit allen Erbern gedencken und den mit Erbern lüten wol ist der Sonne ist ain kungklicher sterne und ain leycht und ouge diser welt und er schinet surch sich selber und erluchtet die andern stern alle und ist under den siben planetten der mihest und zertailt die zit und sinen louff erfuller er in ainem ganzen Jare er machet ouch den menschen wol zu legend an dem libe und daz antlüt machet er im schön und wolgeschaffen mit grossen ougen und mit ainem grossen bart und langes haur und machet den menschen nach der selle von innen glych sinen und machet in nach andrer sachen wyse und daz man

he is thus called Mars because he reigns thus under the seven planets. Thus, there must be much fighting and war in those same years, and when Mars now moves in the Sun's course, then one can seldom see him, yet whenever he reigns, then the masters say that one sees him above the Sun. This thus signifies [a] great defeat against the nobility, [and] thus, that princes and lords, and also knights and squires, should not go to war in that same year because they will be laid low. However, the peasants have good fighting because all things go closely according to their desires. And therefore, the children who are then conceived when Mars reigns, they become quite bellicose people, and do not have the nature with those who are called Sanguine, because they are pugnacious and persevere often and mostly to their battles. However, if one views him [Mars] below the Sun, then he has a nature somewhat like those who are called Melancholic: they are quiet, silent fighters and they succeed well in their wars. And in the year when he reigns, then there a star usually reigns which is called Comet, and in whatever land it is seen, there will be, without doubt, great drought and hunger in that land. Because one cannot see it in all countries, because it is low in the heavens and close to the moon, thus the moon's shadow surrounds it, therefore one can see it well [when] the Sun is in the sign which is called Cancer or Leo, and in whatever year he reigns, then the Sun and the moon of the year are sickly. Whoever is born under the planet, he becomes red with any darkness,[150] like those [who] are burned by the Sun, and also becomes unvirtuously bellicose and likes to foment dissatisfaction among people and has, among the 12 signs, Aries and Scorpio and their temperament and nature. And Mars fulfills its course in five hundred and thirty days.

I, Sun, say to you briefly that my brilliance is above all planets. My rising gives light to the day and my setting draws the stars delicately, and makes the human beings beautiful and lighthearted, which no other planet can do.

The Sun is the fourth planet, and is hot and dry and is joyful, an inflowing light to all those who ever lived on earth. It is a planet beautiful and joyful, and his face lights up human beings, and also illuminates people with all honorable thoughts, and [it] is then well with honorable people. The Sun is a kingly star and a light and eye of this world, and shines in itself and illuminates the other stars in all ways, And [it] is, among the seven planets, the middle one, and divides the time, and completes its course in exactly one year. It also makes human beings better, strengthening them in their bodies, and he makes the face beautiful and beautifully created, with large eyes and with a large beard and long hair, and makes the human beings to resemble their souls, and makes them wise

---

150 Or "unclarity"

40

in gar lieb haut und macht in künstenrich und listig in allen dingen und nachdem planeten sind genattürt die sangwini dann die selben lüte sind gar begriffenlichen in allen dingen und kunsten und sind aber an göttlichen dingen und artikeln gar zwiffelhäftig und sind auch unkunsch und werdent gar lichtillichen erzürnet doch so ist eß und sie bald hin weg daz kinde daz denn da geborn wirt des Jarß so der sonn rengnirt daz wirt flaischhold und ain wysse varbe und mit ain wenig rötte gemischet darumb und nut vil hannß nach der sunnen gelichnusse ud schinet usswendig gar guot und sind doch vast lut nach irem houpte doch maint man daz eß gar wyse lute werden die under dem sonne geboren sind und fröhlich und werdent bosen luten vind der sonne hat under den Siben planeten und under den zwolff zaichen den löwen mit siner Complexion und der nattures und erfüllet sinen louff in einem gantzen umgenden Jare

according to other things and so that one loves them a lot and makes them rich in arts and talents and clever in all things. And the Sanguine are formed like the planet, because the same people are very talented in all things and arts, and are, however, dubious about divine things and items, and are also blindly passionate and often become easily angered, though it also quickly passes away from them. The child that is then born in the year when the Sun reigns, that [child] will have lovely flesh and a white color and with a little red mixed with that, and not a lot of hair, according to the similitude of the Sun. And [they] shine outwardly quite well and are yet loyal people according to their head, yet one opines that they are very wise people who are born under the Sun, and joyful, and become the enemies of evil people. Among the seven planets and among the twelve signs, the Sun has Leo with its temperament and nature, and completes its course in one entire, circumventing year.

145^r **Hie sagt venus von siner nattur und aigenschaft und die under mir geboren werdent die sind nydig und hässig und darzu unkusche**

Here Venus speaks about its nature and character and those born under me are envious and ill-tempered and additionally blindly passionate.

Venus. der planet ist kalt und feucht und gelückhafftig und volbringt iren louff in dryhundert und dry und virtzig tagen Venus ist ain guoter gemainsamer sterne und temperiret marß boßhait und hat ain wolschinende varbe und schinet under dem gestirne dar mitlicllichen und ist anzu sehende alß die sonne und ire kind sind gel und unküsch wenn venuß rengiret so ist gut nüwe claider anlegen

The planet Venus is cold and wet and cheerful, and completes its course in three hundred and forty-three days. Venus is a good, collaborative star and tempers Mars' spite, and has a nicely appearing color, and shines under the stars very lovingly and is to be considered like the Sun, and her children are yellow and blindly passionate. When Venus reigns, it is good to put on new clothes.

Item wenn venuß vor dem sonne gaut so haisset er lutzefer und wann venuß nach dem sonne gaut so haisset venuß vesper venuß machet an dem menschen ain schön person und ouch mit grossen ougen alß der somer anschinende ist und machet den menschen mit der sele wytschwaiffent und an gaistlichen dingen irrig daz sind die dye da haissent Colerici die hand zwyffelhäfftig sinne und belibent doch nit an Item zwyf fel und darumb so sind sie ussgeschiden von denen die da haissent Sangwini wer darunder geboren wirt der wachset nit zu lang mittelmässig und mit grossen ougen und ougbrawen nach dem sonnen alß vor stant undwirt senfftmütig und wolredent und gar züchtig und zucht sich ouch selber gar adenlichen und rainklichen und höret ouch geren saitenspil und tantzet och gerne der planet hat under den zwolff zaichen den ochsen und den libra daz ist die mit iren natur

Item: When Venus precedes the Sun, then it is called Lucifer; and when Venus follows the Sun, then Venus is called Vesper. Venus tidies up human beings to beautiful people, with large eyes as when the summer is shining, and makes humans to be wandering of soul and mistaken in spiritual things. There are those who are called Choleric, they have doubtful minds and yet do not remain with their doubts, and because of that, they are separated from those who are called Sanguine. Whoever is born under [Venus] quickly grows to average height, with large eyes and eyebrows according to the Sun as was written before, and is gentle and well-spoken and quite demure, and also draws to themselves the noble and pure, and also likes to hear string music and likes to dance. The planet has among the twelve signs Taurus and Libra, that is those with their nature.

144^v **Mercurius fürin ist min nattur also betzaichnot min figur mine kind sind hüpschund suptile und waß sie tund daz thund sie in schneller yle**

Mercury: fiery is my nature, thus characterizes my configuration. My children are pretty and subtle and whatever they do, they do that in greater haste.

MErcurius der planet temperiret mit siner natur also komet er zu guoten planeten so ist er ouch guot aber by bösen so ist er auch böß mercurius machet den menschen herlichen an der person und machet den menschen schön doch mit lutzer haarß und wyß nach

The planet Mercury tempers with his nature. Therefore, when he arrives at good planets then he is also good; near evil, however, he is also evil. Mercury makes human beings more splendid to other persons and makes human beings beautiful, yet with little

der sele und suptil und hat och wyshait gat lieb und ist aineß gute siten und ander guten rede gar wolredent doch nit vil und guotes raateß und gewinnet vil fründe Mercurius gaut dem sonne nach und hat ainem schin den man gar selten sicht dann er ist dem sonne gar nsch die under mercurius geboren werden die gewinnend gross zene und sind wyser rede by den lüten und pleich un der varbe und studirend geren und sind still lüte und suptil und stant geren vil an in und hand nit alß boßhait an in selber Mercurius erfullet sinen louff in dreyundert und achtunddryssig tagen Daz sind die melancholici volbringt ir ding haimlicher und rengnirt mit der Junckfrowen und mit dem zwiling under den zwölff zaichen die an den himeln stand

hair, and wise of soul, and subtle, and also loves wisdom very much, and is well-spoken about being well-mannered, and speaking well yet not too much, and about good advice, and gains many friends. Mercury follows the Sun, and has a shine which one sees very seldom because he is close behind the sun. Those who are born under Mercury, they gain large teeth and are speak wisely to the people, and [are] pale in color, and like to study, and are quiet people and subtle, and like to fit in with them, and have nothing of anger in themselves. They are the Melancholics [who] complete their things secretly. Mercury completes its course in three hundred and thirty-eight days, and reigns with Virgo and Gemini among the twelve signs that stand in the heavens.

Hie sagt die mon von iren naturen die sind mengelaye sitenes und spricht mine kind nyemant gern undertänig sind und min figerur nymat allen planetten ire Natur und kain vester wanckel müttige ist denn ich selber zu diser frist

Here speaks the Moon about its natures, which are varied customs, and says "My children do not like to be subservient to anyone, and my configurations take their nature from all planets, and none of the planets is more strongly fickle than I myself at this time."

144ʳ DIe planete die mone Ist die nidorst planet und ist kalt und fücht und auch tugendhaft und ist herre aller füchten dingen und aller schnellest an Irem louff Dann sie louffet in ainem monat also ferre alß der sunn in ainem gantzen iare und alle die die da böse fuchtikeit an inen hand die selben sind ire kind und aller maist so wirt des menschen plut nach dem mone getemperirt und darumb so ist unß gar nütze daz wir wol wissent des moneß ganck und in wolhen zeichen sie gange worum eß ist gar forcklichen wen man Ires louffes nit war nympt Dan sie ist der nidrost planet und zuchet der andern planeten nattur an sich ain michel tail und darumb so müssent wir Iren louff baß wissen wann der andern planeten wann sie alleß daz rengnirt daz in unß ist Dey mone machet den menschen wyt schwaiff also daz er nit mag beliben an ainer statt und machet den menschen under wyle frölich und under trurig also zwayerlay foch day merer tail frölich und machet dem menschen ouch ain krumi nasen und sind füchter natur und haissent flegmatici und sind träg und hand gern ougen also daz ainß grösser ist dann daz ander und luna die mon erfüllet Iren louff alle monat und erlicht die nacht und entlihnet ire liecht von dem sonne und meret und mindert sich von dem sonne also auch hie nach geschriben staut und die kind die sie machet daz werdent gewonlichen knaben und die hand vil gemainsamkait mit den menschen

The Moon is the lowest planet, and is cold and wet and also virtuous, and is lord of all wet things. And [it] is fastest in its course because it moves in one month as far as the Sun in an entire year. And all those who have evil wetness in themselves, the same are its children, and most of all, human blood is thus tempered according to the Moon, and therefore, it is very useful to us that we know well the movement of the Moon and in which sign it moves, because it is quite frightful if one does not perceive its course, because it is the lowest planet and draws the nature of the other planets to itself to a large extent, and therefore, we must know its course better than that of the other planets, because it rules everything that is in us. The Moon makes wanderers of human beings, so that [the human] cannot remain in one place, and makes human beings intermittently joyful and sorrowful again; thus two different things, yet the larger part joyful. And [it] makes a crooked nose for human beings, and [they] are wet by nature and are called Phlegmatic, and are sluggish, and likewise eyes that are also larger than those of others. And Luna, the Moon, completes its course every month and illuminates the night and borrows its light from the Sun, and waxes and wanes from the Sun, as it is also subsequently written here. And the children whom it makes, they are usually boys and they have much in common with human beings. And whenever the Moon reigns, then it is not

143ᵛ und wenn die man regniret so ist nit guot kain ding anfahen Daz lang weren sol alß buwen und solche ding Dann daz ist ain unstäte zyt und ist unbeliblichen waß zu der zit angehaben wirt und die mone machet den menschen plaich under dem antlüt und mit flecken und machet in böß und unsinnig also daz er böß und zornig wirt Daz ist von Ireß wandelß wegen und daß die mon ist in ainem jeglichen zaichen drifthalben tag und haut under Ir den kreps ~

good to start anything that should last long, like building and such things, because that is an uncertain time and whatever is erected at that time is transient. And the Moon makes human beings pale of face and with freckles, and makes him angry and senseless, because he becomes angry and wrathful. That is because of its changeability, and that the Moon is in any one sign for three-and-a-half days and has Cancer under it.

Von den planeten löffe und irer natur und warumb sie iren ganck habent allhie ~

About the courses of the planets and their nature, how and why they have their movements here.

Es ist zu wissent von den siben planeten und von Irer natur daz es got also geordnit haut Der ob dem gestirne ist also wölher planet ainem steren aller nähost gaut von selben steren empfahet er sin natur und sullich stern sind kalter und ettliche nasser nattur ettliche truckner ettlicher haisser nattur die selben naturen zühet der Der mensch von dem gestirne Etliche menschen sind kalt und truckner natur Die selben menschen schwigent garen und die sind ungetrwe menschen Etliche sind kalter und nasser natur die redent vil und sind unverträgenlichen Etliche menschen sind haiss und truckner natur die sind gähmütig und künund hand gern vil wybe und sind doch an der liebe unstätt wölher haisser und truckner natur ist der hat die beste nattur In Im und der ist gern milt und Ergidig und het vast frowen lieb und ist stätt an der liebe

143ʳ und darumb so sagen unß die buoch daz an dem steren den wir haissent marß daz der unlüges pflege wann er ist haisser und kalter natur und trucken dey natturen koment zu der luterkait Die mone ist der aller minst under den sibn planeten und louffet aller nähst by der erden Darumb so richtet sich die welt aller maist nach dem mone Cometa ist ain steren der selbe steren erschinet nymer dann so sich daz rich verwandeln wil den stern sol man kiesen oder ansehn Daz er von dem schin der von Im schinet alß der mone unde der steren louffet nit under andern steren Die buoch sagent unß dat eß am liecht sye Daz got mit sinem gewalt entzündet hab In den lüfften Duch so mainent ettlich daz eß geren tür werde In wolhem lande er gesehn werd ~ ~

It should be known about the seven planets and about their nature, that God, who is above the stars, has thus ordered it that whichever planet approaches a star most closely receives its nature from the same star, and [some] such stars are cold and some wet by nature, some dry [and] some hot by nature—the same natures which the human being draws from the stars. Some human beings are cold and dry by nature; the same human beings are often silent, and they are disloyal human beings. Some are cold and wet of nature; they talk a lot and are quarrelsome. Some human beings are hot and dry of nature; they are undaunting and bold and like many women and are fickle in love. Whoever is of the hot and dry nature, he has the best nature in him and he is gladly patient, and generous, and ambitious, and has a firm love of women and is constant in love. And therefore, the books say to us that, about the star that we call Mars, that it fosters disaster,[151] because it is of hot and cold nature, and dry natures approach purity. The Moon is the absolute lowest among the seven planets and moves closest to the earth; therefore, the world aligns itself mostly according to them Moon.

Comet is a star. The same star never appears, except if the kingdom wants to transform itself. One should select or consider the star, that it shines from the light that [comes] from it, like the Moon, and the star does not move among other stars. The books say to us that it is a light, which God ignited with his power in the heavens. Some also opine that there will be beggars or vagabonds[152] in whichever country it is seen.

Von des sonnen louff wie der louffet durch die zwolff zaichen des himelß tag und nacht

About the Sun's course, how it moves through the twelve signs of the heaven day and night.

WEr recht wölle wissen des morgen louff der muoss des ersten wissen wie der sonne Durch die zwölff zaichen gang und louffet In ainem Jaure durch alle zaichin und belib in einem jeglichen ~~tage~~ zaichin xxx tage wann eß sind zwollff zaichen Arieß thureß Gemini und die andern etc Die zaichin alle durch loufft die mon in ainem monat und durch louffet also alle zaichin in dryssig tagen und belibet un ainem ieglichen zeichen Druhalben tage und durch den louff diser zaichen So komet die mon zu dem zaichin da der sonn inn ist und wirt dem veraint so haisset denn die mon

142ˇ Inbrünstig wenn da verendet sie iren louff wenn aber die mone schaidet von dem sonne xii gräd Daz geburet an den himeln hymilß vahet sie an zulüchtende und

Whoever correctly wants to know the morning course, that one must first of all know how the Dun moves through the twelve signs and moves in one year through all signs, and remains in any one sign 30 days because there are twelve signs: Aries, Taurus, Gemini and the others, etc. The Moon moves through all the signs in one month, and thus moves through all signs in thirty days, and remains in any one sign three-and-a half days. And through the movement of these signs, the Moon then arrives at the sign which the Sun is in and is united there with the same [the Sun]. Then the Moon is called fervent, because it ends its movement. However, when the Moon separates from the Sun by 12 degrees (that is, roughly 61 miles in the heavens), then it begins to shine and is seen by

---

[151] According to Lexer, one of the meanings of *lücke* is *gelücke*, which leads to *ungelücke*, or disaster. Mars promoting "not lies" does not fit the context and would be a convoluted way of promoting the truth, particularly as hot and cold do not lead to positive outcomes.

[152] *Gerentûr* leads nowhere. *Grenter*, on the other hand, leads to "beggers/vagabonds" in Lexer and Grimm. According to Grimm, this is only attested in the western dialects, and points to a French origin of the original text.

wirt gesehen von den menschen und nympt also ir liecht von dem sonne wann sie alle wegen glich in ainer grössin ist aber wenn sie inbrünstig mit dem sonnen ist So mag man man Ir liecht vor Dem sonnen nit gesehen ~

Hie sagt es in wölchem zaichen der sonne In ainem ieglichen monat stand und louffe ~
In dem Jenner ist der sonne in dem zaichen daz da haisset wasser mon und in dem hornung ist der sonn in dem vische und in dem mertzen so ist er in dem wider in dem apperellen so ist er in dem stier in dem mayen so ist er in dem zwiling in dem höwet so ist er in dem löwen In dem ougsten So ist er in dem Junckfrowen In dem Ersten herbst monat so ist er in der wage In dem andern herbst so ist er in dem scorpion In dem ersten winter monat So ist er in dem schützen in dem letsten monat so ist er in dem stainbock.[7]

Hie fähet an ain buoch, und daz da saget wie der lyb innwendig gestalt sye.
Item hie an dem ersten von dem hirn. Daz hiern gyt allen gelidern verstantnuß, wenn die funff sinne des menschen ligent darinne verschlossen.

142ʳ **Das Hertz**
Das hertze gyt allen gelidern werme und pluot, und erneret die sele und behalt daz leben.

Die leber
Die leber git allen gelidern feüchtikait zutrincken, wann sie zücht daz tranck an sich uss dem magen.

Die niern
Daz hirn gyt allen geliden verstanntnuß, und die niern gebent die gepurt, wann die same von allen gelidern in sie komet, und die natur die ain frow zu ainem mann haut, und ain man wider ainer frowen, daz bringt sie an die statt da sich die frucht erhept und da belibet. ~

Item zway löcher gand in den halß, in daz ain gaut die spyz und daz tranck in den magen, und in daz ander gaut der lufft und der autem zu der lungen.
Nun gaut der autem also in die lungen, wann sie ist alß ain plaß palg ob ~~der lungen~~ dem hertzen daz sie den kalten lufft an sich zühet und die hitze mit dem autem wider heruss zühet. Daz loch hat ain uberlid alß man daz essen und daz trincken an sich zücht, so duot sich daz lid zu, also daz die spyß icht da hin yn fare, dann sie höret in den magen und nit
141ᵛ in die lungen. Und wenn der mensch des autemß betarff, so duot sich daz lid uff und dücht den den kalten lufft an sich und zücht ouch wider umb denn haissen lufft her uss, also daz der mensche niht ersticke. Und darumb so ist dem menschen nichtzit

human beings, and thus takes its light from the Sun, because they are always the same in size. However, when it is fervent/luminous with the Sun, then one cannot see its light because of the Sun.

Here it is said in which sign the Sun stands and moves in any one month
In January, the Sun is in the sign called Aquarius; and in February, the sun is in Pisces; and in March, it is in Aries; in April, it is in Taurus; in May, it is in Gemini;[153] in July,[154] then it is in Leo; in August, then it is in Virgo; in the first autumn month, then it is in Libra; in the second autumn month, then it is in Scorpio; in the first winter month, then it is in Sagittarius; in the last month, then it is in Capricorn.[155]

Here begins a book, and that [book] says how the body is internally configured.
Item: Here, firstly, about the brain. The brain gives all limbs understanding because the five senses of the human being are contained therein.

The heart
The heart gives all limbs heat and blood, and nourishes the soul, and maintains life.

The liver
The liver gives all limbs wetness to drink, because it draws the drink to itself from the stomach.

The kidneys
The brain gives all limbs understanding, and the kidneys give birth to it, because the seeds come into them from all limbs, and nature, which holds a woman to a man and a man against a woman, [it] brings them to the point where the fruit rises up and remains there.

Item: Two holes go into the throat, food and drink go into the one [and] into the stomach, and air and breath go into the other [and] to the lungs
Now, the breath goes thus into the lungs, because it is like an inflatable bellows above the heart, that they draw cold air to themselves and heat is drawn back out again with the breath. The hole has a cover: when one draws food and drink into one's self, then the cover closes so that the food cannot go in there, because it belongs in the stomach and not in the lungs, and when the human being needs to breathe, the cover opens and draws the cold air into itself, and contracts again in order to draw the hot air out, that thus the human being doesn't suffocate and because of that, nothing is more mortally damaging to the

---

[153] June and Cancer get skipped.
[154] Literally "hay month"
[155] September, October, November, and December.

schädlicherß zu dem tod, wenn sie pestelentz rengnirt, oder sonst böß fücht nebel oder wetter ist dem böser lufft. Die maister die mainend ouch den den lufft und autem, der von dem menschen komet, besunder siechen menschen, wann eß ist dem menschen nichtzit bessers zu der gesundhait, dann guotten und türren lufft, dann es tött den mensch nichtzit schneller denn bößer lofft, dann ergaut von stund an in alle gelider und verunrainet daz plout und daz hertze inn dem libe. ~ ~

von dem magen
Der mag ist alß ain koch und glich alß ain hafen, darinn die spyse südet und töwet, und ist also aller gelider ain koch und ain knecht, wann er kochet und berait allen gelidern die spyse vor und in das raichet. Die füchte hat er von dem trincken und die hitze von dem hertzen und daz für ouch von der lebern. ~

Maister allmonser spricht in dem buoch daz da haisset panthagin, daz ettliche gelider an dem menschen haiß und trucken sind und etliche kalt und fücht an der Nature. ~

141ʳ **Item so sind daz haisse gelider, als daz hertz und die leber und daz miltz und ouch daz flaysch.**
Item so sind daz die kalten gelider: Daz sind alle die, die nit pluteß an inen hand, alß daz bain und der magen und die tärme und die plaß: und waß wir essent, daz gaut unß alleß in den magen und südet darinn alß in ainem haffen, und dar nach so nympt der magen die spyß und daz tranck alß vil im dann füget und neret sich darvon, und darnach so truckt er daz übrig vin im uß in ainen tarme, der in den magen gaitt und denn so nympt der tarme ouch sin kost darvon, und trucket dem daz ander ouch in ainem andern tarme, und alß die spyß und daz tranck darin komet, so zühet denn die leber daz tranck an sich mit ainem schwaiß, recht alß ain mangnet, der daz ysen an sich zühet, und alß bald daz tranck in die leber komet, so verwandelt eß sich und wirt zu bluot. Die leber zücht ouch daz edlest pluot an sich und neret sich darvon.

Item eß gaut ouch ain grosse auder uß der leber und und alß die ain wenig hin dan komet, so tailt sie sich in zway tail, und die ain gaut uff über sich in die vili der audern, die über alle gelider des menschen gand und neret sich darvon wann daz leben an dem pluot staut, und also duot ouch dem hertzen ain auder mit dem besten bluot, und dar nach so zühet daz die lung des plutes schaum an sich und die

human being—whenever the pestilence reigns, or there is otherwise evil, damp fog or weather—than bad air. The masters, they also include the air and the breath that comes from human beings, especially sick human beings, because there is nothing better for the health of human beings than good and more valuable air. Because nothing kills the human being faster than bad air, because it goes thenceforth into all limbs and impurifies the blood around to the heart in the body.

About the stomach
The stomach is like a cook, and resembles a pot within which the food simmers and digests, and is a cook and a servant for all limbs, because it cooks and prepares the food for all limbs and enriches them. It has wetness from the drinks, and heat from the heart, and fire also from the liver.

In the book that is called *Pantegni*, Master al-Manṣūr says that some limbs in the human being are hot and dry, and some limbs are cold and wet in their nature.[156]

Item: These are the hot limbs, like the heart and the liver, and the spleen, and also the flesh.
Item: These are the cold limbs, that is, all of those that do not have blood on the inside, like the bones, and the stomach, and the intestines, and the bladder. And whatever we eat, that goes into us all, into the stomach, and simmers therein like in a pot, and afterward, the stomach takes the food and the drink (as much as is granted to it) and nourishes itself from that. And afterward, it presses the remainder out of itself into one intestine, which desires that [food] in the stomach, and then the intestine also takes up its food from that, and then presses it for a second time also into a second intestine. And as the food and drink arrive therein, then the liver draws the drink to itself using blood,[157] just like a magnet that draws iron to itself, and as soon as the drink arrives in the liver, then it is converted and becomes blood. The liver also draws the noblest blood to itself and nourishes itself from that.

Item. A large vein[158] also leaves of the liver and as it comes a little distance away, then it divides into two parts, and the one goes up higher into the plurality of the veins that extend across all limbs of the human being, and nourishes it, because life depends on blood, and thus the second vein, which goes lower, also does likewise, and also sends a vein to the heart with the best blood. And afterward, the lungs draw the blood's foam to themselves, and the gall the hot blood, and

---

[156] "al-Mansur" = Manṣūr ibn Ilyās, late 14th C. The book (*tašrīḥi*) *Panthagin* probably relates to *liber pantegni*, a compilation of Greek and Islamic medicine by Constantinus Africanus (11 C). al-Monsur is merely invoked as an important Islamic authority, but the contents clearly refer to Classical medicine.
[157] Or "sweat"
[158] Although *ader* means "artery" and not "vein" in modern German, it seems to have meant "veins" and "nerves" in ENHG. Because the *ader* in the text carry blood, "vein" is used for the translation.

140ᵛ galle daz haisse pluot und daz miltz daz aller beste pluot, und dar nach so sänket sich daz pluot in die audern zu den niern und sühet dar nach dadurch und wirt denn zu harne, und darnach so syhet eß durch claine äderlin alß ain schwaiß in die blaßen.

Item so haut die plauß zway ding, die sie zusamen trucket, also wann die plaaß vol wird so truckent sie die ding von ain ander von der schweere des harneß, und denn so gaut der harne von dem menschen, und dar nach so trucket sich die plaase wider zu, daz der harne nit allewegen von dem menschen gaut, und dar nach so gaut die spyse von ainem tarme in den andern, so lang biß des rainen dingß nichsit mer darinn belibt. Darnach so tribet die natur daz übrig von dem menschen, und dar von wird denn der stuolgang. ~ ~

140ʳ [Blank]

the spleen the very best blood, and afterward the blood collects in the veins to the kidneys, and afterward trickles through that and becomes urine. And afterward, it trickles through small veins like blood[159] into the bladder.

Item: The bladder thus has two things which compress it; thus, whenever the bladder becomes full, the things are pressed apart from one another by the weight of the urine, and then the urine leaves the human being and afterward the bladder retracts again, so that the urine does not always leave the human being. And afterward, the food moves from one intestine into the second until nothing pure remains therein any longer. Afterward, nature drives the remainder out of the human being and from there it becomes the stool.

[159] Or "sweat"

46

1. Hans Talhoffer: Insights into the life of a fencing master in the 15th century

Paul Becker

Of the many fencing masters who taught in the Middle Ages, today we mainly know those who captured their art—or had it captured—in words and images, and only if those works have survived to our time. Little is known about the group of people who taught fencing before the 16th century.

One of the few persons with a significant number of reports to draw on is fencing master Hans Talhoffer. Little research has been done on fencing masters and related professions so far, especially when compared to the wealth of knowledge we have about the nobility. It is therefore difficult to assess the significance of Talhoffer, or his influence on others involved with the fighting arts at his time and in the area where he lived. What sparked my own research into Talhoffer years ago was an article by Paul Sander published in 1902, in which he described in detail the balances and accounting of the city of Nuremberg for the years 1431-1440. As it turned out, Talhoffer made his first appearance among the bills that Sander had made accessible.

In this paper, and drawing on the source material available to me, I will attempt to introduce the person of Hans Talhoffer, putting him into his context and evaluating his importance for modern historiography and the research on pre-modern fighting arts. I will mostly draw on legal documents, correspondences, and invoice notes (and certainly also on the preserved manuscripts of Talhoffer as well as their handwritten copies).

Early career

Hans Talhoffer is known to us today mainly as a fencing master because a significant number of manuscripts on the fighting arts of the 15th-16th century were produced as a consequence of his expertise in this field. He probably lived between 1410 and 1470, and at least five manuscripts related to fencing are currently attributed to him, many copies of which were made in later generations. I will not employ the term "fight book" (*Fechtbuch*) when describing each of his works, since as I will demonstrate below, it does not do them justice.

Figure 1: Hans Talhoffer in 1467.
Legend: Talhoffer, 1467. Munich, Bayerische Staatsbibliothek, Cod. icon 394a, fol. 136ᵛ.

Figure 2: A mounted fighter with sword and crossbow.
Legend: Talhoffer, 1459. Copenhagen, Det Kongelige Bibliotek, Thott 290 2°, fol. 96ʳ.

The earliest record that I was able to find of our Hans Talhoffer dates to the year 1431.[158] The accounting books of the city of Nuremberg mention a "Hans Talhofer" who served the city as a *Reisiger*, i.e. a mounted man-at-arms with his own horse. The city's men-at-arms could perform a variety of services for their masters. They had to perform police duties and military service, but were also employed as messengers and escorts for socially or politically important officials. The city's men-at-arms had to take an oath of service that bound them to it both socially and legally. As our Hans Talhoffer worked in the service of the city of Nuremberg, it is probable that he was well informed about its politics, trade, and society.

We can see how much he was involved in the affairs of the city when looking at the long legal disputes between the citizens of Nuremberg and the nobles of Villenbach. It ran from 1427[159] to 1436[160] and had to be decided by the emperor. Following a number of military conflicts in the 1420s, Emperor Sigismund had forbidden trade with Venice and certain other Italian cities on several occasions. During this time, knights

---

[158] Paul Sander, Die reichsstädtische Haushaltung Nürnbergs, dargestellt auf Grund ihres Zustandes von 1431 bis 1440, Leipzig 1902, p. 158.
[159] Regesta Imperii (RI) XI,2 n. 6900.
[160] Staatsarchiv Nürnberg (StAN) Rst. Nürnberg, Losungamt, 7-farbiges Alphabet, Urkunden 1074.

and lords were allowed to confiscate trade caravans that tried to sidestep these orders.

In spring of 1427, merchants from Nuremberg on their way back from Venice were stopped between Landsberg and Augsburg, where the local knights Hans von Villenbach and Conrad von Magenbuch confiscated their goods in the name of the emperor.[161] In court, the merchants were able to prove that they had left Venice before the trade ban came into force, and subsequently were acquitted by Emperor Sigismund.[162] Hans von Villenbach and Conrat von Magenbuch were therefore ordered by the emperor to return the confiscated goods to the Nuremberg merchants. However, Hans von Villenbach's brother was allegedly murdered in 1429, whereupon Hans declared a feud against the cities of Augsburg and Nuremberg, accusing them of involvement in the murder. The legal dispute is closely connected to the group of people from whom Villenbach had taken the trading goods in 1427.

Under unknown circumstances, Hans von Villenbach managed to capture Hans Talhoffer in 1434, and forced him to make a written confession on 20 March 1434. In it, Talhoffer stated that he knew about the involvement of the city of Nuremberg in the murder of Wilhelm von Villenbach, and gave the names of those Nuremberg citizens who were allegedly involved:

"20 March 1434, Nuremberg: Hanns Talhoffer, brought to Salzburg castle by Hanns von Villenbach and released at the intercession of Hanns der Khuchler[163] and his wife Katharina, swears on the occasion of his release in the presence of Heinrich Pieczenauer, Hanns Czawnru{e}d, Hanns Hartlieb, [and] Joerig Schotl [the] "free judge" (*Freischöffen*), that he had been bribed by Hannsen Siegwein and Jakob Auer of Nuremberg, with Hanns Goldner, Ruedel Frais, Fritz Hiern, Lienhart Lauttrar, the servant of Hanns Lieppach, to murder Wilhelm von Villenbach. The deed was done between [the places] Graisbach and Schainneld by Andre Kraczlan, Fritz Pair, Hanns Pestler, Hanns Ott, Fritz Stumpf, [and] Albert Jud. [Nuremberg] mayors Volkmair and Hans Ortlieb forbade Talhoffer to reveal anything of this. Sealed by Hanns v. Villenbach [and] Hanns Chuchler, the brother of Friedrich Lamppotinger. Witnesses of the sealing: Jörig Prakker, Niklas Gearstetter, Friedrich Gansperger, Stefan Dachsberger, [and] Kristan Sewrsinger."

1434 Mar. 20, Nürnberg, Hanns Talhoffer, der von Hanns von Villenbach auf die Feste Salzburg gesetzt worden und durch Für-

---

[161] See note 159.
[162] Ibid.
[163] The noble Hans Kuchler and his wife seem to have had a close connection to Talhoffer, as they vouch for his release. For more about Hans Kuchler and his wife Katharina see Frank Fürbeth, Johannes Hartlieb: Untersuchungen zu Leben und Werk, Tübingen 1992 p. 50f.

sprache von Hanns den Khuchler und seiner Gemahlin atharina freigelassen worden war, schwört bei seiner Entlassung in Gegenwart von Heinrich Pieczenauer, Hanns Czawnru{e}d, Hanns Hartlieb, Joerig Schotl Freischöffen, daß er von Hannsen Siegwein und Jakob Auer von Nuernberg mit Hanns Goldner, Ruedel Frais, Fritz Hiern, Lienhart Lauttrar, Hanns Lieppachs Knecht bestochen worden, den Wilhelm von Villenbach zuermorden. Ausgeführt ward die Tat vor Andre Kraczlan, Fritz Pair, Hanns Pestler, Hanns Ott, Fritz Stumpf, Albert Jud zwischen Graisbach und Schainneld. Die Bürgermeister Volkmair und Hans Ortlieb verboten Talhoffer davon etwas zu verraten. Siegler: Hanns v. Villenbach, Hanns Chuchler, Friedrich Lamppotingers Bruder. Siegelzeugen: Jörig Prakker, Niklas Gearstetter, Friedrich Gansperger, Stefan Dachberger, Kristan Sewrsinger.[164]

Shortly after his release, on 23 March 1434, Talhoffer wrote to Erhard Haller, his superior in Nuremberg, who was mercenary master and the city's head tax collector (*losunger*).[165] Talhoffer claimed that he had been forced to make this confession, stating that if he had not done it, they simply would have killed him and used someone else they could force to testify. He stated that Hans von Villenbach had taken him to *westvalen*,[166] that is, to the Vehmic court (*Femegericht*)—probably the one in Salzburg—where he had him imprisoned by a judge. Talhoffer then vowed to do everything in his power to rectify the matter. In September 1434, writing from Vienna,[167] he asked Haller for safe conduct and inquired about the right place to answer for his actions and to have his revocation notarised.

He then seems to have gone, via Julbach, to Hals near Passau, where he planned to ambush Hans von Villenbach and take revenge. However, it appears that he was unable to do so. In February 1435,[168] Talhoffer left Hals near Passau via Linz[169] to Vienna, where he waited for further instructions from Haller.

A little later, on 4 March 1435,[170] he sealed his revocation of the forced claims against the Nuremberg citizens. In another document, Thomas Werder, judge of Amstetten (near Ulm), and members of the Amstetten's Market Council sealed Talhoffer's revocation. As both documents are dated 4 March 1435, it can be assumed that Talhoffer went to Amstetten to have his claim recorded at the local court. It is possible that he chose the town because he was considering social and legal

---

[164] Staatsarchiv Nürnberg (StAN) Rst. Nürnberg, Losungamt, 7-farbiges Alphabet, Urkunden 873.
[165] StAN Rst. Nürnberg, Losungamt, 7-farbiges Alphabet, Urkunden 874.
[166] *Westfalen* ("Westphalia") or *westfälisches Gericht* ("Westphalian Court") were common names for the *Fehmegericht* ("Vehmic court" or "secret court") at the time.
[167] StAN Rst. Nürnberg, Losungamt, 7-farbiges Alphabet, Urkunden 898.
[168] StAN Rst. Nürnberg, Losungamt, 7-farbiges Alphabet, Urkunden 928.
[169] StAN Rst. Nürnberg, Losungamt, 7-farbiges Alphabet, Urkunden 930.
[170] StAN Rst. Nürnberg, Losungamt, 7-farbiges Alphabet, Urkunden 938.

Figure 3: How to secure a captive.
Legend: Talhoffer, 1459. Copenhagen, Det Kongelige Bibliotek, Thott 290 2°, fol. 139ʳ.

security for himself there.[171] It was not until August of 1436 that the emperor found Nuremberg innocent of Hans Villen-bacher's charges.[172]

Investigating the lawsuit prompted me to analyse a number of written accounts and the people mentioned in them. This, in turn, provided the chance to examine even more primary records that eventually may help paint a clearer picture of the life of Hans Talhoffer, his environment, and even his personality. Hans Talhoffer was a mounted man-at-arms in the service of the city of Nuremberg starting no later than 1431 and continuing until at least 1435. During this time, he gained a lot of experience with legal procedures in Southern Germany and became acquainted with the lower nobility.

It is not apparent from the sources exactly how long Talhoffer was in the service of the city of Nuremberg. However, we have evidence that he was also in the service of the archbishops of Salzburg, and later probably also the archbishops of Regensburg. Even during his service as a Nuremberg man-at-arms, he probably served the Archbishop of Salzburg in the secret court, but unfortunately the sources do not mention any details. In a document written in 1433, Talhoffer merely stated that he

---

[171] It is hoped that further research will expand our knowledge on Talhoffer's relation to Amstetten, and possibly even his place of origin.
[172] See note 160.

served in the secret court and confirmed that he was compensated for all expenses.[173]

After Talhoffer had managed to withdraw from the lawsuit in 1435 (which was closed just a year later), he seems to have received the office of *Kastner* (steward) in Hohenburg near Amberg, in the service of the bishop Konrad VII of Regensburg. He was apparently appointed in 1437, as he is mentioned in a document written in May of the same year, along with Andre Puntzinger zum Rossstein, who had been appointed *Pfleger* (caretaker) of Hohenburg by the bishop for that year. The offices of *Kastner* and *Pfleger* were usually appointed for a limited period of time, so the officeholders changed frequently. Hans Talhoffer seems to have been in the service of Bishop Konrad VII of Regensburg (†1437) for at least five years, according to a document of his successor bishop Friedrich II of Regensburg in 1444:

> "Bishop Friedrich II[174] of Regensburg admits a debt of 21 pounds of Regensburg pennies to Hanns Talhouer, Kastner of Hohenburg in the Nordgau, resulting from five years of his service to [the bishop's] predecessor in office, bishop Konrad VII of Regensburg. As compensation, [Friedrich II] consigns a house to him in Hohenburg[175] on the Hag, in which he had already lived, and which had previously belonged to Mendorffer."

> *Bischof Friedrich II. von Regensburg verschreibt sich gegenüber Hanns Talhouer, Kastner von Hohemburg auf dem Norgka(e)w über Schulden in Höhe von 21 Pfund Regensburger Pfennigen, die aus fünf Jahren Diensten des letzteren für seinen Amtsvorgänger Bischof Konrad VII. von Regensburg resultieren, u. übergibt ihm als Ausgleich dafür ein Haus das Hohenburg(sic), gelegen auf dem Hag, das der genannte Kastner jetzt schon bewohnt, u. das früher der Mendorffer inne hatte.*[176]

In 1467, 23 years after the house had been bequeathed to him as compensation in 1444, Talhoffer sold it to Bishop Heinrich IV of Regensburg.[177] In the document issued for this purpose, he is referred to as the former Kastner of Hohenburg; records show that he held this office for the last time in 1447.[178] It is also stated that Talhoffer already lived in the house consigned to him. The *Pfleger* at Hohenburg at this time was the brother

---

[173] Urkunde: Salzburg, Erzstift (798-1806) AUR 1433 IV 15.
[174] Friedrich II is a brother of the noble knight Lord Werner von Parsberg, who served in the city of Nuremberg and held the office of Schultheiß (head of the muncipality) in 1442.
[175] Hohenburg near Amberg, in Bavaria, Germany.
[176] Regest zu BayHStA, Hochstift Regensburg Urkunden 1290.
[177] BayHStA, Hochstift Regensburg Urkunden 1565.
[178] BayHStA, Hochstift Regensburg Urkunden 1317.

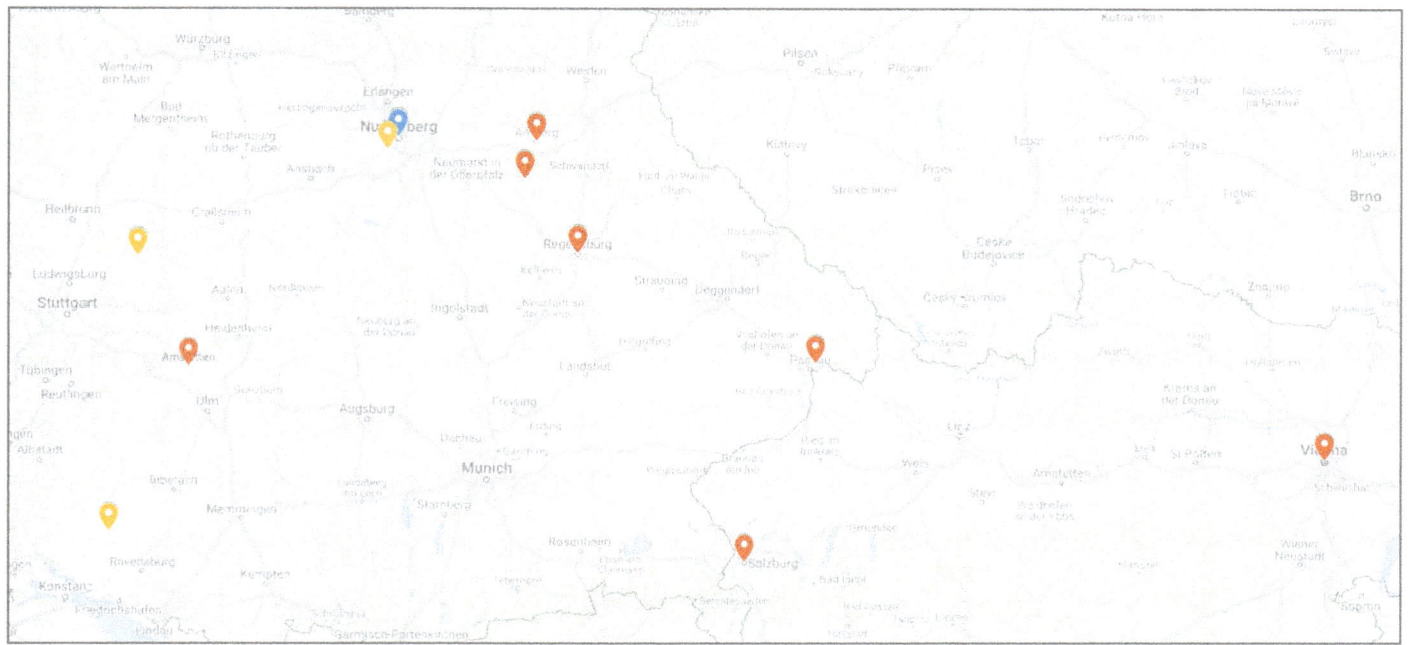

of the bishop of Regensburg, Werner von Parsberg, a highly re-spected nobleman and knight who served the city of Nurem-berg starting 1431 with eight horses,[179] and whom the city sent when dealing with the regular and high nobility. Parsberg ev-entually became mayor of the city in 1442. In the years 1436-1440, the town sent for his services several times, all letters be-ing addressed to Hohenburg.[180] It seems that he, like Talhoffer, had taken up residence there since about 1437-38. The knights and travellers of the city were allowed to stay outside Nurem-berg, as long as they could be in Nuremberg on call within a prescribed time. This special regulation also explains how the man-at-arms Hans Talhoffer could be in Nuremberg and simul-taneously in the service of other lords, such as the bishops of Salzburg or Regensburg.

As *Kastner* of the *Pflege* Hohenburg near Amberg (in the vi-cinity of Nuremberg), he had a fixed salary. In the Duchy of Bavaria at that time, a *Pflege* was an administrative unit or dis-trict. As Kastner, Talhoffer had the task of administering the landlord's income in the *Hofkammer*, i.e. the storage of the sub-jects' payments in kind, as well as supervising the work to be done. Holders of the office of Kastner also often exercised lower jurisdiction. The significance of Talhoffer's office is already ev-ident here, as he appears as a witness in documents of the *Pflege* Hohenburg. This Hans Talhoffer thus held an important and lucrative office. Around 1437, when Talhoffer took up office, the *Pfleger* was Andre *Puntzinger*. He came from a local noble fam-ily, whose members are attested several times in documents of the 14th-16th century in the area of Amberg, where they are

Figure 4: Significant loca-tions in Talhoffer's life. Nuremberg is in blue, the locations of his patrons are in yellow, and locations he travelled to during the Vill-enbach affair are in red. The unlabeled red pin south of Amberg and southeast of Nuremberg is Hohenburg, where Talhoffer lived for much of his life.

---

[179] Sander p. 156.

[180] See R39XIV: 8 ß 2 hl "gen Hohenburg nach Herrn Werner von Parsberg, herzukommen" Sander, p. 563 and also pp. 601, 605, 606 and 610.

mentioned as the holders of important offices.[181] Talhoffer thus worked closely with a very respected family in Bavaria in the *Pflege* Hohenburg.

Whether Hans Talhoffer the *Kastner* is the same person as our fencing master remains unclear until, for example, a seal of the former can be compared with that of the latter. So far, however, the two biographies seem to complement each other. Andre Punzinger, the previous *Pfleger* in charge at Hohenburg, had simultaneously acted as a district judge. It is therefore tempting to assume that our Talhoffer, the fencing instructor, was subsequently appointed because of his experience in judicial matters (gained in Nuremburg and Salzburg) which he would have needed in his new position (but again, this must remain speculation for the time being). At this time, Talhoffer may have established connections to the Lords of Stain via Nuremberg. The following invoice note by the city of Nuremberg for courier services, written in 1438, suggests that the Lords of Stain were already connected with fencing in the city of Nuremberg in that time.

Figure 5: An onomantic diagram and associated text. Legend: Talhoffer, 1448. Gotha, Forschungsbibliothek Erfurt/Gotha, MS Chart.A.558, fol. 13ʳ.

"Sterneck, like Hans von Stein[182] and others of his friends, had written to us because of a [person] here at our place [i.e. Nuremberg] from the companions of [the count] of Oettingen who wanted to learn to fight here."

Sterneck, als uns Hans von Stein und andere seine Freunde geschrieben hatten von eines wegen, der hier bei uns lag in Geleit von des Ottingen Bitte wegen, hier fechten zu lernen[183]

Talhoffer's career as servant at the court had begun early, and by this point, he had acquired further knowledge as a man-at-arms and official. In the 1440s, he seems to have offered this knowledge to other noble lords, which led to his first personal manuscript, dated to 1443.[184] The manuscript, now kept at the Research Library Erfurt/Gotha and recorded as MS Chart. A. 558, clearly was

---

[181] The Puntzinger family were a noble dynasty, known for the bravery of Rüdiger Puntzinger who captured the enemy's flag in the Battle of Gammelsdorf in 1313. The Puntzinger family already owned Rossstein Castle near Amberg and Hohenburg in the early 14th century, where Andre Punzinger resided and named himself after it.

[182] At this time, a certain Hans von Stein held an office at Ronsberg, a castle owned by the baronial vom Stain family. His relation to David und Bubbelin vom Stain, who appear in one of Talhoffer's books, is unclear.

[183] Sander, p. 600.

[184] Yet another year is given in the section with long sword verses (*Zettel*). On fol. 18r it says: *anno dm~ xlviij Jar etc* [...]. This additional date is puzzling. If this '48' is supposed to represent the year 1448, it can be assumed that the creation of this book took a longer time and the *Zettel* of Liechtenauer were not added until 1448.

in Talhoffer's possession and bears his name. On folio 1ʳ, the year 1443 is written together with the ownership entry "This book is [the property] of master Hans Talh..." (*Das buch ist meister hanssen talh...*[185]). It is thus a book that Talhoffer caused to be produced for his own use. In it, he collected, and caused to be written and illustrated, various kinds of knowledge that related to his own life and experience.

The first major part of the manuscript is an onomantic text generally attributed to a Johannes Hartlieb, a doctor working in Vienna in the early 14th century (see fig. 5). Recent research shows that this is probably not, as previously assumed, the Wittelsbach personal physician Johannes Hartlieb, but rather another Johannes Hartlieb who was working in Vienna at the time and wrote or translated several manuscripts for clients.[186] One of his clients is the above-mentioned Hans Kuchler with his wife Katharina, for whom he copied or translated a "Moon Prophecy Book" between 1433 and 1435.[187] This is the same Hans Kuchler and his wife who vouched for Talhoffer's release from prison in Salzburg.[188] The book was also written at the same time that Talhoffer can be traced several times to the Austrian region around Salzburg and Vienna, i.e. around 1433-35. It can thus be assumed that Talhoffer came in contact with the onomancy treatise through his relations with the Kuchlers, possibly while he was in Vienna. It is also interesting that Talhoffer's copy of it includes some words in a typical Austrian dialect.

The onomancy is followed by a copy of the mnemonic verses of Johannes Liechtenauer's art of long sword fighting, mounted combat (see fig. 6),[189] and later in the book also armoured fighting.[190] It was, at the latest, when he began his service as a mercenary in Nuremberg that Talhoffer came into contact with the art of fencing and fencing instruction. It is quite possible that he received fencing lessons here according to Liechtenauer's tradition and therefore wrote down Liechtenauer's verses in his book, copying them from a written model. Even though he does not call himself (nor is called by others) a master of the fighting arts (*Fechtmeister*) or fencing master (*Schirmmeister*), he later presents himself as such and certainly saw comprehensive training.[191]

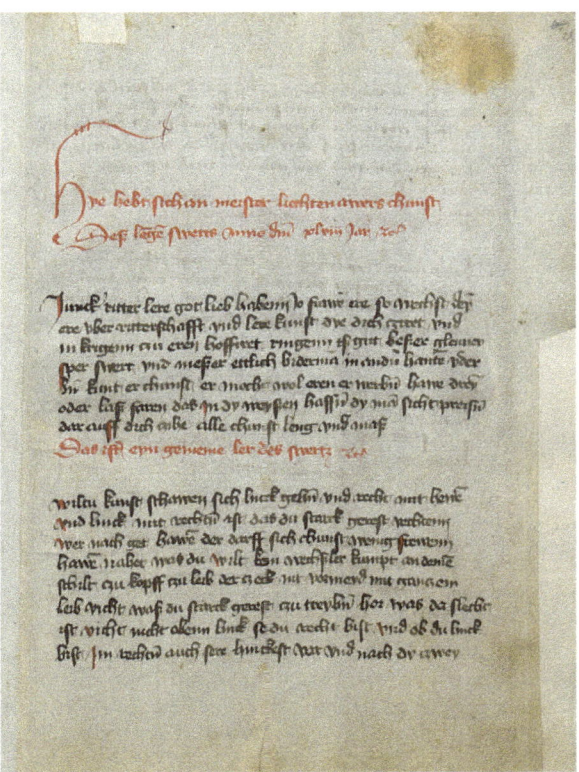

Figure 6: The teachings of Liechtenauer.
Legend: Talhoffer, 1448. Gotha, Forschungsbibliothek Erfurt/Gotha, MS Chart.A. 558, fol. 18ʳ.

---

[185] The remainder of the text is missing.
[186] For the various people from the 1430s with the name Johannes Hartlieb see Fürbeth, Hartlieb, p. 30 ff.
[187] Fürbeth, Hartlieb, p. 50 ff.
[188] See note 163.
[189] MS Chart.A.558, fol. 18r. ff
[190] MS Chart.A.558, fol. 48v.
[191] MS Chart.A.558, fol. 25v. ff Talhoffer stages himself as a fencing master from the news to the court battle.

Possibly from 1435 onwards, the eventual fencing master Paulus Kal was also active in Nuremberg; in 1435-39, the Nuremberg accounts mention a gunmaster named Paul.[192] He could well be identical with the fencing master Paulus Kal, as can be seen from his professional career.[193] Paulus Kal is mentioned in the service of Nuremberg and, from 1450 onward, in the service of the Wittelsbach dynasty as a gunsmith and *Schirmmeister*.[194] It is therefore probable that the Nuremberg gunsmith Paul from the second half of the 1430s was the selfsame fencing master Paulus Kal. The records also mention a fencing master Hans Lecküchner as a Nuremberg citizen; his art is contained in two manuscripts, from 1478 and 1482.[195] Fighting, fencing (*schirmen*), and Liechtenauer's tradition can therefore be considered already well-established in the city of Nuremberg in Talhoffer's time.

Figure 7: Trial by combat with clubs and shields. Legend: Talhoffer, 1448. Gotha, Forschungsbibliothek Erfurt/Gotha, MS Chart.A. 558, fol. 42ʳ.

The fact that even the distant Lords of Stain and their friends had showed interest in fencing lessons in Nuremberg around 1438 shows that Hans Talhoffer may have had contact with Liechtenauer's teaching during his time in Nuremberg. One could come to this assumption because Talhoffer copied the *Zettel* of Liechtenauer in his first work and only in this one, and Nuremberg is the first station in his career known to us. The fellowship of Liechtenauer may also have already been present in Nuremberg at this time, as I later show using the example of the fencing master Paulus Kal, who was in the direct lineage of the Liechtenauer tradition and lived in Nuremberg that time.

Following Liechtenauer's teachings in MS Chart. A. 558 are drawings of clothing and arms for the use in judicial duels, followed, in turn, by a detailed and captioned series of

[192] Sander, p. 465.

[193] The *Schirmmeister* Paulus Kal appears in several documents (charters, letters and account books), initially in relation to Nuremberg. He committed a breach of peace there in 1449, probably by drawing a weapon, and exchanged letters with the city in the same year. After entering the service of Duke George IX of Bavaria in 1450, he is mentioned, among other things, as the commander of 12 arquebusiers. A good summary of primary sources related to Paulus Kal can be found here: https://talhoffer.wordpress.com/2011/07/03/paulus-kal-a-schirrmeister/ (accessed on 1 November2020).

[194] A certain "master Paul fencing master" (*maister Pauls Schirmaister*) is mentioned as a gunmater in Nuremberg's defence regulation issued in 1449, see G. Franz, Quellen und Erörterungen zur bayerischen und deutschen Geschichte, vol. 8 1860, p. 202. He appears in several letters received by the city, e.g. in one where Duke Louis of Bavaria wants to admit him into his service; see Dieter Rübsamen, Das Briefeingangregister des Nürnberger Rates für die Jahre 1449–1457, Nürnberg 1997, no. 1796. For more sources, see again https://talhoffer.wordpress.com/2011/07/03/paulus-kal-a-schirrmeister/ (accessed 1 November 2020)

[195] "These are master Hans Lecküchner of Nuremberg's art and verses that he had made and rhymed himself for [fighting with] the Messer" (*Das ist herrn hansen Lecküchnerß von Nûrnberg künst und zedel ym messer dy er selbs gemacht und getichtt hatt* [...]). Universitätsbibliothek Heidelberg, Cod.Pal.germ.430, fol. 2r.

images showing the announcement of the duel,[196] the training of the fighters, the duel it-self (see fig. 7), and eventually the death of one of them. It is the most detailed representation of this process known to me, and the level of detail may be the reason for the many copies of this particular manuscript that were made. Six copies of this manuscript are known to exist,[197] making it the most often-copied medieval fight book of any master. In this work, Talhoffer probably recorded the knowledge and ex-perience he gained in his service at the judicial courts of Nuremberg, Salzburg, and Regens-burg. This suggests that he might have already been offering his services as a fencing instruc-tor at that time, as he explicitly presents him-self as a fencing instructor.[198]

The judicial duel with club and shield is fol-lowed by depictions of armoured fighting with the short sword (see fig. 8), as was typical of affairs of honour, i.e. fights from wantonness (*muotwill*) among the wealthier social classes.

We also encounter here the first recording of the wrestling teachings of Master Ott. Unlike the fight scenes depicted earlier in the manuscript, the wrestling illustrations are equipped with short written descriptions. The final part of the manuscript contains drawings of military equipment; they are probably the result of Talhoffer's accumulated knowledge in this respect from his time in Nuremberg. In those years, the city of Nuremberg had a well-equipped armoury and a well-organised city defence, as we can see in the written accounts.[199] As a servant and man-at-arms of the city of Nuremberg, can be assumed that Talhoffer knew the arsenal and all buildings, warehouses, and equipment belonging to the city defence.

The contents of the manuscript MS Chart. A. 558 correspond to what we know of Talhoffer's life between c. 1431 and 1447. In the manuscript, he does not emphasise the execution of fighting techniques, but instead showcases his knowledge and skills so that he can offer his services to lords and knights.

Considered together with the list of his previous employers, he has thus built up considerable experience (and references)

Figure 8: Procession into the list for an armoured duel. Legend: Talhoffer, 1448. Gotha, Forschungsbibliothek Erfurt/Gotha, MS Chart.A. 558, fol. 55ᵛ.

---

[196] Judicial duels were fights that were fought before court of combat (germ. *Kamfgericht*) based on local law. For examples of the procedures of regional judicial duels and courts of combat see Friedrich Majer, Geschichte der Ordalien, insbesondere der gerichtlichen Zweikämpfe in Deutschland, Ein Bruchstück aus der Geschichte und den Alterthümern der deutschen Gerichtsverfassung, Jena 1795 p. 294ff.

[197] (1) Cod.Guelf.125.16.Extrav., Herzog August Bibliothek Wolfenbüttel, Germany (2) 2° Col. MS. philos. 61, Universitätsbibliothek Göttingen, Göttingen, Germany (3) 2° MS iurid. 29, Universitätsbibliothek Kassel, Kassel, Germany (4) MS 26.236, Metropolitan Museum of Art New York City, USA (5) MS 014, Kunstsammlungen der Veste Coburg, Coburg, Germany (6) Cod.icon. 395, Bayerische Staatsbibliothek, Munich, Germany.

[198] See note 191.

[199] See Sander, Rechnungswesen p. 466.

da stat her dauid vnd buppellm vom stain gebrueder vn(d) hand die kunst die jn disem büch stat gelernnot von hansen dalhofer.

Figure 9: David and Bubbelin vom Stain with Talhoffer. Legend: Talhoffer, 1440s-50s. Berlin, Stiftung Preußischer Kulturbesitz, MS 78.A.5, fol. 63ʳ.

until the year 1447. In fact, his services are manifold: serving at regional courts, instructing and coaching the fighters of judicial duels, serving as a mounted man-at-arms or mercenary—Talhoffer would have been a valuable employee for high lords and dignitaries. To round it off, he was also well connected in the areas of Nuremberg and Regensburg, and had met a number of respected nobles.

Later career

Unfortunately, we lack archival documents or letters to help describe the rest of Talhoffer's life as vividly. We now have to rely almost exclusively on his other books, which he mainly wrote for various noblemen. His first customers were probably the lords David and Bubbelin vom Stain,[200] who had already sought fencing instruction in Nuremberg in 1438. He probably taught them himself, because Talhoffer later mentions himself in the book as the teacher of the two (see fig. 9).[201]

In this work he describes course of several duels and fighting techniques with different weapons in captioned images. However, the emphasis here is on the presentation of fencing techniques and the corresponding equipment in both text and illustrations. Therefore, the book is both a showpiece and an aide-mémoire for the lords vom Stein, its images and brief descriptions allowing them to recapitulate what Talhoffer had taught them. This manuscript is thus his first real "fight book" that he produced for a customer or employer.

Probably because of his services to the lords vom Stein, Talhoffer established himself in Swabia, for he produced another work similar to this one for the young noblemen (*Junker*) Leuthold von Königsegg (also Swabian). The latter is attested for the years 1446 to 1473.[202] In this second 'proper' fight-book, Talhoffer again names himself as the teacher of the Junker (see fig. 10): "Here Leuthold of Königsegg wants to learn [to

---

[200] The fight book manuscript for the lords vom Stain is kept at the Stiftung Preußischer Kulturbesitz in Berlin, Kupferstichkabinett, manuscript 78 A 15.

[201] The inscription reads: "There stand lord[s] David and Bubbelin vom Stain, brothers, who have learned the art contained in this book from Hans Talhoffer." (*Da stat her dauid vnd buppellm vom stain gebrueder vn(d) hand die kunst die in disem büch stat gelernndt von hansen dalhofer.*) Stiftung Preußischer Kulturbesitz in Berlin, Kupferstichkabinett, manuscript 78 A 15, fol. 63r.

[202] Luitild sealed a document on 23 July 1454, see Landesarchiv Baden-Württemberg, Abt. Hauptstaatsarchiv Stuttgart, H 52 a U 240.

fight] in earnest from Talhoffer [...]" (*Hie will lẅtold von küngsegg lernen zů dem Ernst von dem Talhofer [...]*).[203]

The manuscript's structure is similar to that for the lords of Stain. In this work, however, Talhoffer sets forth an instruction concerning the duties in the relationship between the fencing master and the young nobleman. He also wrote down his own mnemonic verses on fencing, which he probably composed himself. Only afterwards we find a series of captioned illustrations, which show techniques of different fighting styles and a duel of the Junker. When exactly Talhoffer was in the service of the Junker and had this book made for him is not clear from the sources that I know.

Talhoffer's own progress can also be seen in this book, as he seems to have considered his art so important that he reminded the Leutold—in written form!—to keep what he had learned a secret. The use of his own fencing verses at the beginning are also a novelty, as he still used Liechtenauer's Zettel in MS Chart. A. 558, and none at all in the book for the lords vom Stain.

The most impressive of Hans Talhoffer's manuscripts in terms of content and illustration, and succeeding the first opus MS Chart. A. 558, is manuscript Thott.290.2°, which was written around 1459. One could argue that this was his masterpiece, in which he combined much of the knowledge he had gathered in his professional career. He developed in it almost all the areas of specialised knowledge that had been present already in his previous works. In particular, the first sections Talhoffer added with his fencing poems and instructions for fighting respectively judicial duels[204] are carried out more comprehensively than in his earlier works. The course of the judicial duel, however, is not shown, in contrast to MS Chart. A. 558. The depiction of Talhoffer in person that can be found in the latter[205] is roughly on par in terms of quality with all of the images in Thott. 290. 2°. Since he had his portrait drawn in both manuscripts, even the significant difference in age is visible (see fig. 11).[206] Both MS Chart. A. 558 and Thott. 290. 2° seem to have been produced for Talhoffer himself, which were intended to serve as references of his craft for potential employers, without betraying too much knowledge in the text and the images.

Figure 10: Talhoffer instructs Leuthold von Königsegg. Legend: Talhoffer, 1440s-50s. Königseggwald, Königsegg-Aulendorf Collection, MS XIX.17-3, fol. 25ʳ.

---

[203] MS XIX.17-3, Königsegg-Aulendorf Collection, Königseggwald, Germany, fol. 1v.
[204] MS Thott.290.2°, fol. 1r-10v.
[205] MS Chart.A.558, fol. 25v.
[206] MS Thott.290.2°, fol. 101v.

Figure 11: Illustrations of an older (left) and younger (right) Talhoffer.
Legend: Talhoffer, 1459. Copenhagen, Det Kongelige Bibliotek, Thott 290 2°, fol. 101ᵛ. Talhoffer, 1448. Gotha, Forschungsbibliothek Erfurt/ Gotha, MS Chart.A.558, fol. 25ᵛ.

The last known work of Talhoffer, and thus probably the last piece of evidence of his biography, is the manuscript Cod. icon. 394a from c. 1467.[207] He probably had it made for Count Eberhard im Barte of Württemberg, as suggested by the coat of arms and the year written underneath (see fig. 12).[208] Unlike all other of his manuscripts, this work is limited to fencing techniques with very short verses. The manuscript contains neither a preface nor a fencing recital. The client had presumably just wanted an illustrated fight book, and even Talhoffer himself was mentioned only briefly. There is a small illustration of him holding a banner (see fig. 1) on which is written "Hans Talhoffer has offered this book and posed [for its images] at Mallen" (*Das buch hatt angeben hans talhoffer vnd gestehen zu Mallen*).

In the same year, in 1467, he sold his house at Hohenburg,[209] which he had received in 1444 from the bishop of Regensburg as compensation for his services. According to the respective deed he had already lived in it before. This retreat or possibly even homestead he now gave up, and there is no information on what could have caused this step. Still, Talhoffer sold his

---

[207] Bayerische Staatsbibliothek München, Cod. icon. 394a, Munich, Germany.

[208] Ibid., fol. 16v.

[209] "Hans Talhoffer, former *Kastner* at Hohenburg in the Nordgau, sells the rights to a house in Hohenburg, which he had received from Friedrich [II of Parsberg], bishop of Regensburg, to Heinrich [IV of Absberg], and relinquishes all his claims to the bishop that result from his office as *Kastner*." (*Hanns Talhover, ehemaliger Kastner in Hohemburg auf dem Norkaw, verkauft Heinrich [IV. von Absberg], Bischof von Regensburg, die Rechte über ein Haus in Hohenburg, die ihm Friedrich [II. von Parsberg], Bischof von Regensburg, verbrieft hatte und verzichtet gegenüber dem Bischof auf alle weiteren Ansprüche aus seiner Tätigkeit als Kastner.*) BayHStA, Hochstift Regensburg Urkunden 1565.

Uff dem anbinden aber ain
vahen mit gewalt.

1 2 6 ∧

house in Franconia in the very same year in which he entered the services of Count Eberhard of Württemberg. It can therefore be assumed that he moved his residence to Swabia where he had already worked for the families of vom Stain and Königsegg. A book of accounts from the court of Count Eberhard provides yet another piece of evidence, incidentally from that very same year.[210] "The Talhoffer" is here given ten *gulden*, three buckets of rye, and fifteen buckets of oats. While this could also be our Hans Talhoffer, there was also a silk embroiderer named Hans Talhoffer working in Stuttgart at the same time, and he appears in local sources until 1508.[211]

At this point, we run out of evidence concerning the life of Hans Talhoffer. We do not know when he died, and it might be misleading to assume that this happened shortly after the last bit of information that we have. 36 years of his life, from 1431 to 1467, can be reconstructed with some certainty, and he is clearly rather old-looking on his portrait in manuscript Thott. 290. 2° from 1459. What happened before and after this timeframe remains unknown to us, at least for the time being.

Figure 12: The coat of arms of Eberhard im Barte. Legend: Talhoffer, 1467. Munich, Bayerische Staatsbibliothek, Cod. icon 394a, fol. 16ᵛ.

---

[210] A 602 Württembergische Regesten/1301-1500, Hausarchiv, Eberhard V. (als Herzog I.) Bestellsignatur A 602 Nr 286 = WR 286.
[211] A 602 Württembergische Regesten/1301-1500, Kanzlei, Quittungen Bestellsignatur A 602 Nr 3596 = WR 3596.

Figure 13: Salzburg Erzstift (798-1806) AUR 1433 IV 15.

Figure 14: Comparing the arms of Thott, Gotha, and Kal. Legend: Talhoffer, 1459. Copenhagen, Det Kongelige Bibliotek, Thott 290 2°, fol. 101ᵛ. Talhoffer, 1448. Gotha, Forschungsbibliothek Erfurt/Gotha, MS Chart.A.558, fol. 28ʳ. Kal, 1470. Bologna, Biblioteca Universitaria, MS 1825, fol. 5ᵛ.

Seal and heraldry of Hans Talhoffer

Seals of Talhoffer are preserved in several of the extant letters and documents. Comparing them allows us to determine whether all of the Talhoffers are, in fact, the same person. The seals which I have been able to assess suggest[212] that the Hans Talhoffer who worked in Nuremberg (see fig. 13) and the one who served in Salzburg are identical. Although the seals are in bad condition, they seem to show the same coat of arms a crown in a shield with maybe swords behind. Unfortunately, I was unable to examine another seal on a document[213] by Talhoffer for the archbishopric of Regensburg to compare it before completing work on this essay. It therefore has yet to be determined if the gaps in Talhoffer's biography can be filled in this regard.

In contrast, we find multiple depictions of Talhoffer's coat of arms in his books. The most detailed one is to be found in MS Thott. 290. 2° from around 1459. Talhoffer's coat of arms consists here of a crown through which crossed swords are thrust on a black background; the crest has a ragged mantling in black and silver together with what appears to be an upright riding crop on top. The motif with crown and swords can already be found in his first work, MS Chart. A. 558,[214]

---

[212] Seals that I have been able to see or compare so far are those on the correspondence between Talhoffer and the city of Nuremberg and the seal on the document for the Hereditary Bishop of Salzburg. See Salzburg Erzstift (798-1806) AUR 1433 IV 15, for Nuremberg StAN Rst. Nürnberg, Losungamt, 7-farbiges Alphabet, Urkunden 874, 928, 938. I was able see the seal of the Salzburg document on a high-resolution picture.

[213] BayHStA, Hochstift Regensburg Urkunden 1565.

[214] MS Chart.A.558, fol. 28r.

where he presents himself as the fencing master in a series of images depicting the chronological order of the ordeal. There and in the self-presentation in MS Thott. 290. 2° fol. 101ᵛ Talhoffer wears a necklace with his coat of arms and a hat, which he decorates with an emblem of St. Mark in MS Chart. A. 558 and St. John in MS Thott. 290. 2°.

The coat of arms on fol. 28ʳ in MS Chart. A. 558 is depicted next to the illustration of a person preparing for a fight with club and shield according to Frankish judicial customs. Presumably this is Talhoffer himself, too, but the original coat of arms was painted over. The crown-and-crossed-swords motif is preserved, but the background was painted over with blue and red (colours also found in the ragged mantling). Likewise, the crest with the riding crop was painted over and now shows an anchor with the letter S wrapped around its shank, the latter being exactly where the former crop was, and the lines of the crop's whipcord still visible. [215]

Figure 15: Defaced pendant of the fencing master.
Legend: Talhoffer, 1448. Gotha, Forschungsbibliothek Erfurt/Gotha, MS Chart.A.558, fol. 33ᵛ.

However, not only the coat of arms was altered, but also Talhoffer's necklace that showed his coat of arms was later erased on each occasion (see fig. 15). It can thus be assumed that a later owner of the book wanted to remove all traces of the former owner in order to use it for their own purposes. This also explains why on fol. 1ʳ, Talhoffer's ownership entry and the image above (possibly his coat of arms, too) were painted over with glue and chalk and so made unrecognisable.

From this crest one could assume that Paulus Kal later came into possession of the book, because his coat of arms was, indeed, an anchor with a sword thrust through the upper loop of its shank, and an S for *Schirmmeister*. Perhaps he was not able or willing to fully redesign the shield, so that he merely altered Talhoffer's coat of arms and identified himself with the help of the repainted crest. This, along with the other erasures and images painted over, would mean that he wanted to "cover up" all that connected him to Talhoffer.

Conclusion

If we look at the periods of Hans Talhoffer's life that are known to us, the image of a man appears who did not consider himself a fencing master first and foremost, nor one who was primarily occupied with teaching the art of fighting. In contrast to Paulus Kal, Talhoffer was not called a *Schirmmeister* or *Fechtmeister* in letters or documents, whereas Paulus is given these attributes in the majority of sources on him, and even includes the S for

---

[215] Paulus Kal's coat of arms can be seen in his fighting book MS 1825, Biblioteca Universitaria di Bologna, Bologna, Italy fol. 5v and in a later copy of the 16th century MS Chart.B.1021, Forschungsbibliothek Erfurt/Gotha Gotha, Germany fol. 1v. I have copies of his seal on two documents Tiroler Landesarchive TLA Urkunde I 2743 and Urkunde I 4793. Both seals are identical to the coat of arms in the books. In the seals are the letters "P.S." are incorporated. As he starts his documents with "*Ich pauls koll schirmaister*[...]" the "P.S." in the seals could mean "*Pauls Schirmaister*".

*Figure 16: Seals of Paulus Kal.
Legend: See note 215.*

Schirmmeister in his seal.[216] Talhoffer primarily was a mounted man-at-arms who built up a good reputation through various services and acquired a wide range of knowledge which made him the person of choice for several employers.

His most important set of knowledge and skills, however, related to the art of fencing. His familiarity with judicial duels is reflected in his fight books as well as other documents. His commissioned works for noblemen always included fighting techniques for duels. These techniques, related types of clothing and weapons, the duel itself, its rules, and the required preparation for the fight are at the heart of these manuscripts. However, they are much different from the descriptions of a comprehensive and complete fencing doctrine, as we find it in "typical" fight books, for example by Paulus Kal, some of which also contained glosses on the Liechtenauer doctrine.[217]

In his own books, Talhoffer added much more varied content that reflected his entire portfolio, such as extensive illustrated knowledge of intelligence service (such as knot writing), war equipment, Hebrew, astrology or onomancy. In MS Thott. 290. 2°, he also offers us his own take on mnemonic

---

[216] See note 215.

[217] Apart from the manuscript Cgm 1507 (Bayerische Staatsbibliothek Munich, Germany), the other manuscripts of Paulus Kal—besides their extensive series of illustrated fighting techniques—always contained a comprehensive textual part with the recital of and the glosses to the fighting art of the "Fellowship" (*Gesellschaft*) surrounding Master Johannes Liechtenauer. While this textual part is preserved in the Viennese manuscript MS KK5126, (Kunsthistorisches Museum Vienna, Austria) from fol. 104v onwards, the most text pages in manuscript 1825 (Biblioteca Universitaria di Bologna, Bologna, Italy) are lost but the text of the teachings of Liechtenauer starts on fol. 45r. On fol. 45v the text ends in the middle of the verse at the end of the page. It can therefore be assumed that the rest of Lichtenauer's teachings once followed here, as in the Viennese Paulus Kal manuscript MS KK5126 (Kunsthistorisches Museum Vienna, Austria).

verses on fencing. Talhoffer thus gives us an impressive and unique insight into the life of a man who taught and created fencing in the 15th century.

For posterity, Hans Talhoffer's writings were and are of great importance, not so much because of the actual instructions on fighting, but rather with regard to legal and social history. The courts of combat played a prominent role in this. How important the works of Hans Talhoffer were can easily be gauged. It can be shown that for the copyists of the 16th to 18th centuries, the legal aspect of Talhoffer's works was probably particularly important; most of the scenes of procedure for judicial duels were copied from the manuscript MS Chart. A. 558 and inserted into collective manuscripts on the subject of law, as can be seen in the copies in manuscripts MS 014 (Art Collections of the Coburg Fortress, Coburg, Germany), 2° MS iurid. 29, (University Library Kassel, Kassel, Germany) and Cod. Guelf. 125. 16. Extrav. (Herzog August Bibliothek, Wolfenbüttel, Germany). All three manuscripts place the copy of the scenery between legal texts, often from Würzburg, and also contain a written description of the martial court in Franconia, which is from 1432-1459 based on the Würzburg arch chamberlain Kraft Zobel (von Giebelstadt) mentioned there and can be traced to a written source from the first half of the 15th century.

The manuscript 2° MS iurid. 29 from Kassel is particularly striking. Its texts refer to Würzburg, the archbishops and several Würzburg legal texts. The manuscript MS Chart. A. 558 was copied at least five times,[218] possibly because of its detailed depiction of the judicial duel. Thus it is probably the most copied or partially copied medieval fighting manuscript.

Especially scholars from the fields of legal and social history would benefit from a deeper analysis of the texts and illustrations in Talhoffer's manuscripts. This is not to say, of course, that it was not valuable also for other disciplines. Although we do not find description of any comprehensive fighting doctrine, research on and reconstructions of the historical fighting arts can also benefit to no small degree from an examination of these books. Even historians of technology and fashion will find ample evidence in Talhoffer's multi-content manuscripts.

In conclusion, it should be pointed out that apart from Paulus Kal, no other fencing master has been reflected so often in the written and pictorial sources of the Middle Ages and the Early Modern Era as Hans Talhoffer. This makes the study of his person indispensable for modern research on this topic.

---

[218] See note 197.

Bibliography

Primary sources

List of abbreviations

BayHStA   Bayrisches Hauptstaatsarchiv
OeStA/HHStA   Österreichisches Staatsarchiv, Haus-, Hof-
                 und Staatsarchiv
StAN   Staatsarchiv Nürnberg
TLA   Tiroler Landesarchiv

Documents and invoices

Nürnberg, Staatsarchiv Nürnberg, Rst. Nürnberg, Losungamt,
    7-farbiges Alphabet, Urkunden.
München, Bayrisches Hauptstaatsarchiv, Hochstift Regens-
    burg, Urkunden.
Stuttgart, Landesarchiv Baden-Württemberg, Abt.
    Hauptstaatsarchiv Stuttgart, H 52 a Archivalien aus dem
    Germanischen Nationalmuseum in Nürnberg / 1341-1834 .
Stuttgart, Landesarchiv Baden-Württemberg, Abt.
    Hauptstaatsarchiv Stuttgart, A 602 Württembergische
    Regesten/1301-1500.
Innsbruck, Tiroler Landesarchive, Urkundenreihe I (TLA Ur-
    kunde I).
Wien, Österreichisches Staatsarchiv, Haus-. Hof- und Staatsar-
    chiv, Allgemeine Urkundenreihe (AUR).

Manuscripts

Gotha, Forschungsbibliothek Erfurt/Gotha, MS Chart.A.558.
Gotha, Forschungsbibliothek Erfurt/Gotha, MS Chart.B.1021.
Heidelberg, Universitätsbibliothek, Cod.Pal.germ.430.
Wolfenbüttel, Herzog August Bibliothek Wolfenbüttel,
    Cod.Guelf.125.16.Extrav.
Göttingen, Universitätsbibliothek, 2° Col. MS. philos. 61.
Kassel, Universitätsbibliothek, 2° MS iurid. 29.
New York City, Metropolitan Museum of Art, MS 26.236.
Coburg, Kunstsammlungen der Veste Coburg, MS 014.
Munich, Bayerische Staatsbibliothek, Cod.icon. 394a.
Munich, Bayerische Staatsbibliothek, Cod.icon. 395.
Munich, Bayerische Staatsbibliothek, Cgm 1507.
Königseggwald, Königsegg-Aulendorf Collection, MS XIX.17-3.
Berlin, Stiftung Preußischer Kulturbesitz, Kupferstichkabinett,
    MS 78 A 15.
Copenhagen, Det Kongelige Bibliotek, MS Thott.290.2°.
Wien, Kunsthistorisches Museum, MS KK5126.
Bologna, Biblioteca Universitaria, MS 1825.

Printed sources

Historische Kommission bei der Königl. Bayerischen Akademie der Wissenschaften, Quellen und Erörterungen zur bayerischen und deutschen Geschichte, vol. 8, München 1860.

Secondary literature

Friedrich Majer, Geschichte der Ordalien, insbesondere der gerichtlichen Zweikämpfe in Deutschland, Ein Bruchstück aus der Geschichte und den Alterthümern der deutschen Gerichtsverfassung, Jena 1795.
Fürbeth, Frank, Johannes Hartlieb: Untersuchungen zu Leben und Werk, Tübingen 1992.
Sander, Paul Die reichsstädtische Haushaltung Nürnbergs, dargestellt auf Grund ihres Zustandes von 1431 bis 1440, Leipzig 1902.

68

2. Talhoffer galore

Dierk Hagedorn

In this article, I want to touch on two major subjects: the first is to highlight the tradition line in which fencing master Hans Talhoffer lived and worked, while the second is to give a brief overview of the fencing treatises that are connected to his name. For this paper, I have re-used some passages from my essay "German *Fechtbücher* from the Middle Ages to the Renaissance".[219]

A brief history of books

The two doyens of research as far as the German sources are concerned, MARTIN WIERSCHIN[220] and HANS-PETER HILS,[221] have both published extensive volumes about the German medieval fencing literature. In their works, they list and describe a considerable number of fencing manuscripts, called fight books (*Fechtbücher*, in German). Over the past decades, numerous fencing books have been rediscovered or found that were believed lost or were previously unknown. MARTIN WIERSCHIN listed 47 manuscripts in 1965, and twenty years later, we find 55 treatises in HANS-PETER HILS' catalogue. Speaking of catalogues, the *Katalog der deutschsprachigen Handschriften des Mittelalters*[222] from 2009 offers 47 again—but these are only the illustrated ones. At the time of writing, 74 fight book manuscripts from the German speaking regions are known.

The earliest known fight book worldwide is German, known by its signature *I.33*,[223] which dates to the beginning of the 14th century. It is kept in the Royal Armouries in Leeds and is—despite its origin—written in Latin, the language of scholarship of the time, the language of the church. And indeed, in this book we meet a priest who teaches his student the art of fencing (see fig. 1). We do not know the reason for this. Was fencing a popular pastime among the clergy? Was he a younger son of the nobility trained in the martial arts who had to seek his career as a clergyman? Was such teaching among monks the exception or the

Figure 1: Priestly fencing with sword and buckler.
Legend: Anonymous, 1320s. Leeds, Royal Armouries, MS I.33, fol. 9ʳ.

---

[219] HAGEDORN (2016).
[220] WIERSCHIN (1965).
[221] HILS (1985).
[222] LENG et al. (2008).
[223] Leeds, *I.33*.

Figure 2: Sword and buckler fencing from Gladiatoria. Legend: Gladiatoria, 1440s. Crakow, Biblioteka Jagiellońska, Ms. Germ. quart. 2020, fol. 54ᵛ.

rule? The Cracow manuscript of *Gladiatoria* (see fig. 2),[224] which on fol. 54ᵛ speaks of the *chuttennische puckler*, or "the buckler of the monks wearing the habit" (*Kutte*), provides only a brief reference.

The following two or three centuries see a surprising number of hand-written fight books—in the German-speaking regions, that is. With their total of 74 known manuscripts from this period, the German-speaking regions of Europe were far more prolific than any other area.

There was a certain transitional period from the hand-written codices to printed volumes. The development was a slow but steady process. The year 1452, in which Johannes Gutenberg began to work on his Bible, can be regarded as the beginning of the art of printing—and coincidentally, this was the same year in which one of the most extensive German manuscripts was completed, the Peter von Danzig manuscript.[225] Nevertheless, it took a considerable amount of time until printing finally took over. The first printed German fight book appears almost three-quarters of a century later (Andre Paurenfeindt's book, published in 1516).[226] Only about a dozen fight books were printed in the 16th century, but still some 30 manuscripts were produced. But when we talk about manuscripts, we cannot ignore one overshadowing figure, a fencing master who appears either by name or by his teachings in 27 out of 74 manuscripts: one cannot write about German fight books without paying tribute to grandmaster Liechtenauer.

He who must be named

Much ink—real and digital—has been spent on the shadowy figure of the medieval fencing master Johannes Liechtenauer, yet still we know virtually nothing about the man. We can obtain only very scarce biographical data from an anonymous codex from Nuremberg[227] that mentions him. On fol. 13ᵛ, we learn that Master Liechtenauer did not in fact invent the art of swordsmanship, but he travelled far and wide in order to further his knowledge:

---

[224] Cracow, *Ms. Germ. Quart. 16.*
[225] Rome, *44 A 8.*
[226] PAURENFEINDT (1516).
[227] Nuremberg, *Cod. Hs. 3227a.*

And above all things shall you note and know that there is only one art of the sword which may have been invented and devised many hundred years ago. It is a foundation and core of the entire art of fencing, and Master Liechtenauer knew it entirely and perfectly and was wholly capable of it. However, he did not invent it himself, as is written above, instead he has travelled and visited many a country because of that true and sincere art, since he wanted to get to know it thoroughly.

Nevertheless, the codex does not offer us certain dates from Master Liechtenauer's life. So far, no reliable biographical evidence whatsoever has been found. His origin remains mysterious, though his name itself may refer to his birthplace. Unfortunately, there are too many towns called Liechtenau or Lichtenau for this to narrow it down much.

Although the person himself remains obscure, Liechtenauer's influence was quite long-lasting. The manuscript from Nuremberg probably offers us the first encounter with the art according to Liechtenauer. A living tradition of roughly 250 years ensued.

Although Liechtenauer's influence was (and among modern practitioners still is) huge, his teachings are actually fairly limited in scope. He has only left us a couple of verses that rhyme more or less elegantly. These verses are obscure and elusive; and without proper explanation they remain almost opaque gibberish and are thus next to useless for any practical applica-

Figure 3: Johannes Liechtenauer.
Legend: Anonymous, 1452. Rome, Accademia Nazionale dei Lincei, Cod. 44 A 8, fol. 2ᵛ.

tion. Luckily, we gain much more insight into Liechtenauer's teachings through secondary sources. The master bequeathed his obscure poems to his pupils and thus to posterity—possibly only orally, possibly already in written form. After all, there may be a reason why they are called *Zedel* in the old sources: "*Zedel*, Latin *schedula*, usually signifies an ephemeral way of notation on single leaves, in contrast to the more durable form of books."[228] Without comments, these verses are difficult or even impossible to understand. According to the writings of his successors and/or students, Liechtenauer produced them in this way in order to prevent all the world from being able to understand these skills:

Therefore he had them written down in hidden and secret words so that not everybody shall hear and understand them. He has done so because of the careless fencing masters who don't

---

[228] MÜLLER (1992), p. 256.

value their art. Thus he wants to avoid these masters either making his art public or spreading it.[229]

Fortunately for us, Liechtenauer's students didn't care about that very much, for in the very next sentence of these introductory words it says:

And these very same hidden and secret words of the notes (Zedel) are written hereafter in the commentaries explained and interpreted so that everybody, who is already able to fence, can read and understand them.[230]

So, although Master Liechtenauer himself wished to hide his knowledge from a broader audience, his disobedient pupils did not share his methods—or rather, they did. We must not forget that the books that contain Liechtenauer's knowledge are manuscripts—one-off items that were not widely available but rather aimed at a single client in order to serve as a memory aid.

Master Liechtenauer has spawned at least one follow-up generation of masters. These are known as the *Gesellschaft Liechtenauers* (Liechtenauer Society) as the name appears in Paulus Kal's manuscripts, as codified for instance in his manuscript from Munich (see fig. 4):[231]

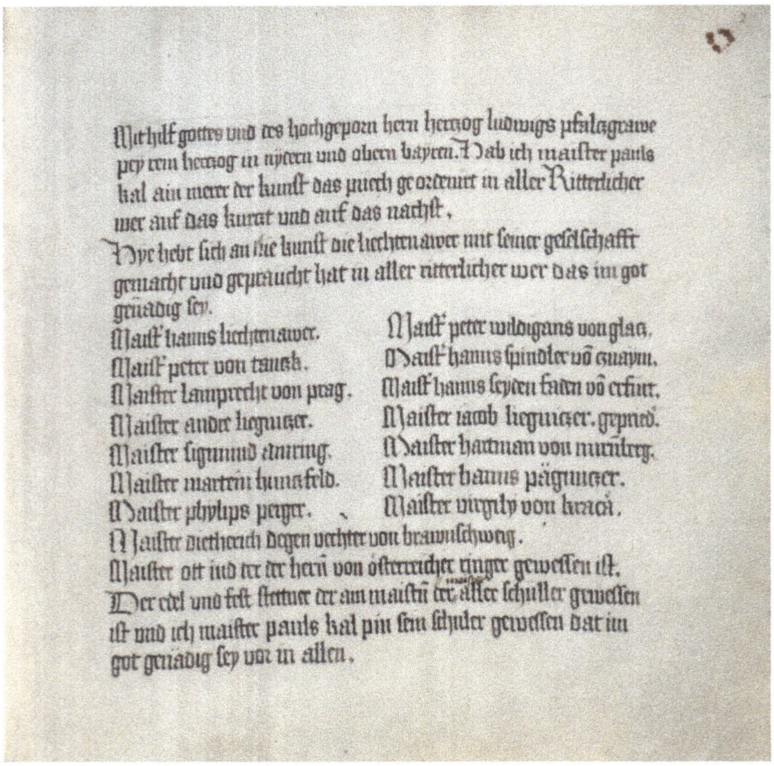
Mit hilf gottes und des hochgeporn herrn herzog ludwigs pfalzgrave pey rein herzog in nidern und obern bayern. Hab ich maister pauls kal ain maister der kunst das puech geordenirt in aller Ritterlicher wer auf das kurrat und auf das nachst.
Hye hebt sich an die kunst die liechtenawer mit seiner gesellschafft gemacht und gepraucht hat in aller ritterlicher wer das im got genädig sey.
Maist hanns liechtenawer. | Maist peter wildigans von glacz.
Maist peter von tanczk. | Maist hanns spindler vō znaym.
Maister lamprecht von prag. | Maist hanns seyden faden vō erfurt.
Maister andre liegniczer. | Maister iacob liegniczer. geprued.
Maister sigmund amring. | Maister hartman von nurnberg.
Maister martein hunczfeld. | Maister hanns pägniczer.
Maister phylips perger. | Maister virgily von kracā.
Maister dietherich degen vechter von brawnschweig.
Maister ott iud der der herrā von österreicher ringer gewessen ist.
Der edel und fest stettner der ain maistā der aller schuller gewessen ist und ich maister pauls kal pin sein schuler gewessen dat im got genädig sey und in allen.

Figure 4: The Liechtenauer Society.
Legend: Munich, Bayerische Staatsbibliothek, Cgm 1507, fol. 2ʳ.

Master Johannes Liechtenauer,
Master Peter Wildigans from Glaz,
Master Peter from Danzig,
Master Hans Spindler from Znaim,
Master Lamprecht from Prague,
Master Hans Seidenfaden from Erfurt,
Master Andre Liegnitzer,
Master Jacob Liegnitzer (brothers),
Master Sigmund Amring,
Master Hartmann from Nuremberg,
Master Martin Huntfelt,
Master Hans Pegnitzer,
Master Philip Perger,
Master Virgili from Cracow,
Master Dietrich, dagger fencer from Brunswick,
Master Ott, the Jew, who was the wrestler of the lords of Austria,
the noble and strong Stettner, who above all was the master of all pupils,
and I, Master Paulus Kal, was his

[229] Rome, *44 A 8*, fol. 9v.
[230] Ibid., fol. 9v.
[231] Munich, *Cgm 1507*, fol. 2r.

pupil, and may God have mercy upon him.

This is not necessarily a formal society of individuals that met on a regular basis; it can rather be seen as a list of masters that were in one way or the other associated with Liechtenauer, the grandmaster. Paulus Kal lists many masters we have references to in other manuscripts, such as Peter von Danzig, Andre Lignitzer and others; some on the list have left no trace, like Peter Wildigans; some may be identical to authors that have bequeathed material to us: Sigmund Amring for instance is certainly identical to Sigmund Emring[232] and Sigmund ain Ringeck;[233] but it is a curious circumstance that there is at least one master contemporary to Kal who is not mentioned: Hans Talhoffer, the fencing master from whom we have the most surviving fight books by far.

The man and his manuscripts

Only scarce biographical material exists about Hans Talhoffer. Even his name appears in various forms: Talhoffer,[234] Dalhofer[235] or Talhofer.[236] Although some recent publications[237] deal with Master Talhoffer, the person behind the manuscripts remains mostly elusive. Most of the obtainable data can be gathered from online sources:[238]

The first time we encounter Talhoffer in historical records is in 1433, when he served Johann II von Reisberg, archbishop of Salzburg, in a Vehmic court. A year later, during or after his service for the archbishop, he was accused of murdering Wilhelm von Villenbach. Talhoffer confessed to having been hired to abduct Hans von Villenbach, the deceased's brother, but testified that others committed the murder. The trial caused some upheaval, but Talhoffer remained in the archbishop's service.

We meet Talhoffer again in the records of the city of Zurich in 1454, when he supervised a public fencing event and had to intervene in a fight that got out of hand.[239]

Talhoffer continued to serve various patrons, such as Leutold von Königsegg, the brothers David and Buppelin vom Stain and Eberhard I, Duke of Württemberg. An extensive manuscript was produced for each of these noble employers, and it is a curious circumstance that both the Königsegg and the Stain version contain the identical illustrated duel—thus it is doubtful whether either duel actually took place or whether the pictures

---

[232] Glasgow, *E.1939.65.341.*

[233] Dresden, *Mscr. Dresd. C 487.*

[234] K, p. 47, 49, 71, 121; M, fol. 136v; the abbreviations are explained in the next section.

[235] B, fol. 63r.

[236] K, p. 121; C, fol. 2r, 101v, 103v.

[237] BURKART (2014), ISRAEL (2017).

[238] wiktenauer.com/wiki/Hans_Talhoffer; talhoffer.wordpress.com/category/a-life-like-that-of-talhoffer/ biography/

[239] HILS (1985), p. 176.

Figure 5: Talhoffer's heraldry. Legend: Talhoffer, 1459. Copenhagen, Det Kongelige Bibliotek, Thott 290 2°, fol. 102ʳ.

were created for advertorial reasons or as a sign of homage to the principal. While in Württemberg's service, his name is recorded for the last time when Eberhard paid Talhoffer ten guilders and some rye and oats.

Up to this day, there is speculation whether Talhoffer was a member of the Marxbrüder fencing guild (brotherhood of St. Mark) since a lion, their heraldic animal, appears in his coat of arms in one of his manuscripts (see fig. 5).[240] Nevertheless, no conclusive evidence has been found to date.

Most fight books are compendia with various sections, sometimes written by a single person such as Hans Talhoffer, sometimes a compilation of works from various masters. These anthologies differ in extent, and the individual components are not identical from one volume to another. Some authors, such as Paulus Kal or Jörg Wilhalm, commissioned copies of their own works more or less unaltered; in contrast, Hans Talhoffer's books vary considerably. We have five archetypical manuscripts from his lifetime, plus one contemporary copy, plus some nine later copies. This makes a total of fifteen Talhoffer-related manuscripts.

---

[240] Copenhagen, *Thott 290 2°*, fol. 102r.

Abbreviations

A: Augsburg, Universitätsbibliothek, *Cod. I.6.2°.1.*

B: Berlin, Kuperstichkabinett der Stiftung Preußischer Kulturbesitz, *78 A 15.*

Co: Coburg, Kunstsammlungender Veste Coburg, *Hz. 014.*

C: Copenhagen, Det Kongelige Bibliotek, *Thott 290 2°.*

G: Gotha, Forschungsbibliothek Schloss Friedenstein, *Ms. Chart. A558.*

Gö: Göttingen, Niedersächische Staats- und Universitätsbibliothek, *2° Cod. Ms. philos. 61.*

Ka: Kassel, Landesbibliothek, *2° Ms. iurid. 29.*

K: Königseggwald, Gräfliches Schloss, *Hs. XIX, 17-3.*

M: Munich, BayerischeStaatsbibliothek, *Cod. icon 394a.*

M2: Munich, Bayerische Staatsbibliothek, *Cod. icon 394.*

M3: Munich, BayerischeStaatsbibliothek, *Cod. icon 395.*

N: NewYork, Metropolitan Museum of Art, *26.236.*

P: Private collection, olim: Vienna, Österreichische Nationalbibliothek, *Cod. Vindob. Ser. Nov. 2978.*

V: Vienna, Kunsthistorisches Museum, *KK 5342.*

W: Wolfenbüttel, Herzog August-Bibliothek, *Cod. Guelf. 125. 16 Extrav.*

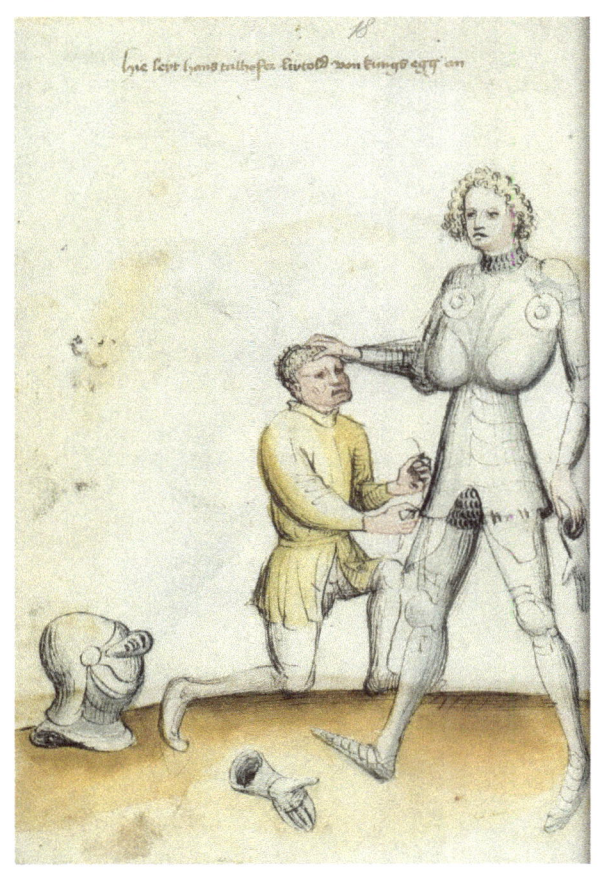

Figure 6: Talhoffer helps a student arm himself. Legend: Talhoffer, 1440s-50s. Königseggwald, Königsegg-Aulendorf Collection, MS XIX.17-3, fol. 9ᵛ.

Though Talhoffer did not personally scribe any of the manuscripts, he claims to have posed for at least two of them, namely C and M. He is also shown helping his patrons don their suits of armour in both K and B (see fig. 6). ERIC BURKART,[241] however, doubts that any of these images can be assigned the status of a true portrait.

To explore thoroughly the contents of each of these manuscripts is beyond the scope of this paper. Much information, however, can be gathered from HILS (1983), LENG (2008) and BURKART (2014). From these sources, we can assume for sure that Talhoffer was a man of multiple interests: apart from fencing in various disciplines, he offers valuable information on why and how a judicial duel was to be held and fought; some of his books contain portions of war books with siege engines and war machines; and he also looked into onomatomantia (working out who will be victorious in a fight, based on his name and the time at which it is to happen).

In order to shed some light onto the web of the connections of the various manuscripts, I shall identify which are the archetypes, which is to say the original version of the text, and which are the copies. Archetypes are set in

---

[241] BURKART (2014).

bold, corresponding copies are indented. Subsequently, however, I will look at the archetypes in more detail.

Archetypes and copies

G	**Gotha**, 1443/48	
	Ka	Kassel (partially), 17th c.
	W	Wolfenbüttel (partially), late 17th c.
	Co	Coburg (partially), 18th c.
	N	New York, 17th c.
	Gö	Göttingen, late 17th c.
	M3	Munich, *ca.* 1820
K	**Königseggwald**, middle of 15th c.	
	V	Vienna, end of 15th c.
	A	Augsburg, before 1561
B	**Berlin**, before 1459	
C	**Copenhagen**, 1459	
M	**Munich**, 1467	
	P	Private collection, 16th c.
	Ka	Kassel (partially), 17th c.
	W	Wolfenbüttel (partially), late 17th c.
	Co	Coburg (partially), 18th c.
	G	Göttingen, late 17th c.
	M2	Munich, ca. 1820

Figure 7: Counter to the "scissors".
Legend: Talhoffer, 1448. Gotha, Forschungsbibliothek Erfurt/ Gotha, MS Chart.A.558, fol. 66ᵛ.

G is an anthology of many disciplines and a collection of various artists. At least four scribes and 11 artists were involved in the process of creating the codex.[242] The entire manuscript has a rather heterogenous appearance and may have been compiled from various sources and then bound into one single volume, a procedure which was common in the Middle Ages and the Renaissance.

- Unarmoured longsword – illustrated: fol. 2ʳ-6ʳ (blank pages: 2ᵛ, 3ᵛ, 5ᵛ, 6ᵛ)
- Johannes Hartlieb's onomatomantia: fol. 7ʳ-17ʳ (blank pages: 12ᵛ, 17ᵛ)
- Master Liechtenauer's teachings for longsword and mounted combat: fol. 18ʳ-23ʳ (blank pages: 21ʳ, 23ᵛ)
- Duel with large shields – illustrated: fol. 24ʳ-48ʳ (blank pages: 42ᵛ, 43ᵛ, 44ᵛ, 45ᵛ, 46ᵛ, 47ᵛ; fol 24ʳ with a pencil sketch; 41ᵛ traced-through sketch from overleaf)
- Harness fencing (Liechtenauer's verses): fol. 48ᵛ

[242] According to LENG (2008).

76

- Harness fencing – illustrated: fol. 49^r-72^r (all verso pages are blank)
- Harness fencing – illustrated (outline sketches, no captions): fol. 73^v- 80^v (all recto pages are blank; 81^r-82 blank)
- Dagger – illustrated: fol. 83^r-104^v (fol. 105^r-109^r blank)
- Ott's wrestling: fol. 109^v-114^v (fol. 115^r-116^r blank)
- Wrestling – illustrated: fol. 116^v-133^v (fol. 134-140 blank)
- War book – illustrated: fol. 141^r-148^r
- Ammunition (17th c. [?] addition)—illustrated: fol. 148^v

K has been in the possession of the König-segg family from its inception to the present. It describes the training of squire Leutold von Königsegg and a stylised duel with an unknown adversary. The codex has spawned two copies, V and A. V is so similar to K that it is not unlikely that both manuscripts stem from the same workshop or even the same artist. The images and the handwriting, however, are a bit less elaborate in V. According to Leng, the watermarks indicate a slightly later date, *ca.* 1480-1500.[243] Altogether, K is the more complete version since V lacks several captions, particularly at the end of the codex. A is an artistically rougher copy, likely of K, not of V, since it contains those captions not included in V.

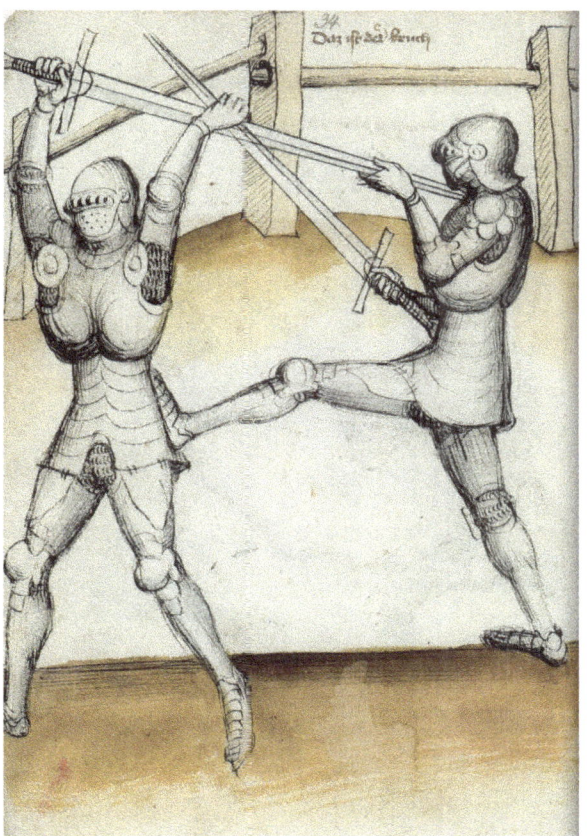

Figure 8: Counter to the "scissors".
Legend: Talhoffer, 1440s-50s. Königseggwald, Königsegg-Aulendorf Collection, MS XIX. 17-3, fol. 17^v.

- Introduction and verses: pp. 1-2
- Stances for harness fencing: pp. 3-17 (some pages are blank: 4, 6, 8)
- Leutold's duel in armour: pp. 18-48 (some pages are blank: 46, 48)
- Dagger: pp. 49-70
- Wrestling: pp. 71-88
- Spear: pp. 89-98
- Fighting on horseback (lance): pp. 99-100, 107
- Fighting on horseback (sword against lance): pp. 101-106
- Fighting on horseback (sword): pp. 109-112
- Fighting on horseback (wrestling): pp. 113-119
- Jumping into the saddle: p. 120
- End page: p. 121

---

[243] Leng (2008), p. 58.

Figure 9: Counter to the "scissors".
Legend: Talhoffer, 1440s-50s. Berlin, Stiftung Preußischer Kulturbesitz, MS 78.A.5, fol. 27^v.

B offers an armoured combat scene in the beginning that is distinctly different from K. This is followed by the same armoured, un-armoured and mounted techniques we are familiar with from K, although here the sequence is torn apart (due to rebinding the manuscript). But while it is Leutold von Königsegg who is being taught by master Talhoffer in K, it is here either David or Buppelin vom Stain. Additionally, the various heraldic shields are different. The overall style of the illustrations is rather similar to K, although some of them, particularly in the dagger and wrestling sections, look much more refined. Some additional material concerning how to dress for a duel and the appropriate weapons is also provided, which is absent from its predecessor, and we are introduced to fencing with the *langes Messer* and the buckler as well as with the flail. A section with the large duelling shields, following Franconian and Swabian law, is also included. In summary, B is a heavily extended version of K.

- Harness fencing: fol. 1^r-27^v
- Fighting on horseback: fol. 28^r-32^r
- Wrestling: fol. 32^v, 38^{rv}, 43^r-44^v, 46^r, 48^{rv}, 51^r-53^v, 57^r-59^v, 61^r
- Staff/spear/flail: fol. 33^{rv}, 45^{rv}, 55^r-56^v, 60^{rv}
- Dagger: fol. 36^v, 39^r-42^v, 46^v-47^v, 49^r-50^v
- Dressing the fighter: fol. 37^{rv}
- Buckler: fol. 54^{rv}
- Weapons for the fight: fol. 62^{rv}
- Brothers vom Stain, Talhoffer: fol. 62^r
- Long shield: fol. 62^v-77^v

C constitutes the next step in the evolutionary process. Talhoffer has—allegedly—not only posed in person for the illustrations, he expands the scope considerably: A lengthy introduction, a war book and sections about cryptography and the Hebrew alphabet are included, as well as chapters about the planets and anatomy. The fight book part has also been extended by fighting with the pollaxe and by duels between man and woman.

- Introduction and verses: fol. 1^r-10^v (6 and 7 are blank leaves)
- Illustrations about cryptography (?): fol. 11^{rv}

- Bellifortis (war book): fol. 12^r-48^v
- Wrestling: fol. 49^r-60^v, 138^r-139^r (139^v is blank)
- Dagger: fol. 61^r-71^r
- Pollaxe: 71^v-74^v, 131^r-137^v
- Spear against sword and other weapons: fol. 75^r-78^v
- Sword against mace: fol. 79^{rv}
- Man against woman: fol. 80^r-84^r
- Harness fencing: fol. 84^v-94^r (from 89^r onwards shown without armour)
- Fighting on horseback: fol. 94^v-97^r, 124^r-130^r
- Duelling shield and sword: fol. 97^v-99^r, 102^v, 116^v-117^r
- Duelling shield and club: fol. 99^v-101^r, 103^r, 110^v-111^r, 116^r
- Image and crest of Hans Talhoffer: fol. 101^v-102^r
- Colophon: fol. 103^v
- Arms: fol. 104^r-110^r
- Duelling shield: fol. 111^v-115^v
- Sword/*langes Messer* and buckler: fol. 117^v-123^v
- Tournament scene: fol. 130^v

The following leaves are inscribed upside down, starting at the last leaf; therefore, the foliation is reversed:

Figure 10: Counter to the "scissors".
Legend: Talhoffer, 1459. Copenhagen, Det Kongelige Bibliotek, Thott 290 2°, fol. 92^r.

Figure 11: Counter to the "scissors".
Legend: Talhoffer, 1467. Munich, Bayerische Staatsbibliothek, Cod. icon. 394a, fol. 31ʳ.

- Cryptography/Hebrew alphabet: fol. 150ᵛ-149ᵛ (149ʳ is blank)
- Planets: fol. 148ᵛ-142ᵛ
- Anatomy: 142ᵛ-140ᵛ (140ʳ is blank)

M, finally, is stripped again of the non-fencing-related material. The chapters are much more coherent than in the previous books, and for the first time we find a whole chapter about true unarmoured combat with the sword. Before, it was mostly armoured combat depicted sans armour—for whatever reason, possibly convenience on behalf of the models. The artistic standard, certainly enhanced in C, reaches its peak in this volume with its slender and elegant late Gothic figures.

Apart from general bibliographical information such as the material (parchment) and the size (333 × 215 mm), LENG[244] relates that four artists were responsible for the images of the codex, with I and II being hard to distinguish—I: 2ʳ-39ᵛ, 40ᵛ-52ʳ; II: 40ʳ, 86ᵛ-109ʳ, 110ᵛ-112ʳ; III: 52ᵛ-53ʳ, 109ᵛ-110ʳ, 113ʳ-122ʳ, 127ʳ-136ᵛ; VI: 53ᵛ-86ʳ, 122ᵛ-126ᵛ.

---

[244] LENG (2008), pp. 54-6.

- Sword (unarmoured): fol. 2^r-17^v
- Halfsword (armoured techniques performed out of armour): fol. 18^r-35^r
- Harness fencing: fol. 35^v-40^v
- Pollaxe: fol. 41^r-53^r
- Duelling shield and club: fol. 53^v-64^v (65^r is blank)
- Duelling shield and sword: fol. 65^v-76^v
- Duelling shield: fol. 77^r-86^r
- Dagger: fol. 86^v-96^r
- Wrestling: fol. 96^v-112^r (112^v is blank)
- *Langes Messer:* fol. 113^r-116^v
- Sword and buckler: fol. 117^r-121^r
- One against two: fol. 121^v-122^r
- Man against woman: fol. 122^v-126^v
- Fighting on horseback (sword): fol. 127^r-131^v
- Fighting on horseback (wrestling): fol. 132^r-133^v
- Fighting on horseback (lance): fol. 134^{rv}
- Fighting on horseback (crossbow against lance): fol. 135^r-136^v

Owner or author

For many scholars, it is a mystery why Paulus Kal omitted Talhoffer from Liechtenauer's society. Hans-Peter Hils suggested that this omission was inspired by envy on Kal's side.[245] I find this unlikely since Paulus Kal found himself in frequent employment by dukes and other noblemen, whereas Talhoffer seems to have been more or less a journeyman in fencing affairs. But what if Talhoffer had never been a member of this illustrious circle anyway?[246]

This suggestion is largely based on the fact that Liechtenauer's verses appear in G. But if Talhoffer merely acquired the book without being its author, this of course does not mean that he must have been a member of this society. It is possible that a work from Liechtenauer or his circle may have come into his hands as an outsider. Of course, this does not exclude that he found such an authority as Liechtenauer interesting, or that he possibly admired or even revered him. Perhaps he even took over, adapted, or plagiarised parts of Liechtenauer's teaching. After all, a lot of Liechtenauer-related terminology has been preserved in Talhoffer's treatises.

But as the anonymous first glossator on Liechtenauer's verses said, the revered master had already travelled to many a country in order to learn and perfect his skills. There may have been fencing terms before Liechtenauer—and parallel to him. After all, there is also a whole chapter in the early Nuremberg fight

---

[245] HILS (1983), p. 112; HILS (1985), pp. 179-80.
[246] As already suggested by MÜLLER (1992), p. 262.

Figure 14 (op, bottom): Preparing to parry a thrown spear.
Legend: Talhoffer, 1448. Gotha, Forschungsbibliothek Erfurt/Gotha, MS Chart.A.558, fol. 56r.

book about techniques of other masters, such as the priest Hanko Döbringer, Andre the Jew, Jost from the river Neiße and Niclas from Prussia. Talhoffer could therefore have learned the material that goes back to Liechtenauer on his own, in order to modify and design it according to his taste or his own understanding of art. After all, he had travelled widely, and his experience was considerable.

So we need to take Talhoffer's authorship of G with a considerable grain of salt. The most obvious circumstantial evidence is his name on fol. 1ʳ. In this regard, Hans-Peter Hils says:

Figure 12: Heraldry similar to Talhoffer's.
Legend: Talhoffer, 1448. Gotha, Forschungsbibliothek Erfurt/Gotha, MS Chart.A.558, fol. 28ʳ.

The manuscript is Talhoffer's oldest known creation and presumably his first work. It was apparently at least temporarily in his possession, indicated by a painted text banner on fol. 1ʳ (today only legible if the page is held against the light): "Daz bůch ist maister hanssen talhoffersz". The last three letters are destroyed by worm damage but can be completed according to the (possessive) genitive 'hanssen' and the same formulation in another manuscript of the Master—Thott 290 2°, fol. 103ᵛ.

That, however, does not qualify whatsoever as a proof of authorship. Another fight book author was Paulus Hector Mair, treasurer of the city of Augsburg. Although he was indeed the patron of three luxuriously illustrated fight books,[247] several additional manuscripts also bear his name—as owner, not as author.[248]

Apart from the name on the first page of G, there is one more important clue that Talhoffer at least owned the manuscript—leaving it an open question nevertheless whether authorship can be attributed to him: a coat of arms is displayed on fol. 28ʳ (see fig. 12)which is similar though not identical to examples from K, B, and C (see fig. 5). Curiously enough, the jousting helmet of the coat of

---

[247] Dresden, Sächsische Landesbibliothek, *Mscr. Dresd. C 93/94*; Munich, Bayerische Staatsbibliothek, *Cod. icon. 393* (1 and 2); Vienna, Österreichische Nationalbibliothek, *Cod. 10825/10826*.

[248] *Cod. I.6.2°.1* – see abbreviation A; *Cod. I.6.2°.2* – a compilation of Jörg Wilhalm's longsword treatise and Lienhart Sollinger's manuscript version of Andre Paurenfeindt's printed fight book from 1516; *Cod. I.6.2°.3* – Jörg Wilhalm's fencing in harness and on horseback; *Cod. I.6.2°.5* – a compilation of records of the Marxbrüder fencing guild, a glossed version of Johannes Liechtenauer's longsword verses and engravings of Maarten van Heemskerck's fencing and wrestling; *Cod. I.6.4°.2* – the so-called Codex Wallerstein or Baumann's fight book, a compilation of armed and unarmed fighting disciplines with detailed descriptions, and an older, uncaptioned part that is connected to the Gladiatoria group of fencing treatises; *Cod. I.6.4°.5* – a draft version of *Cod. I.6.2°.3*; *MS E.1939.65.354* – Gregor Erhart's fight book; *Schätze 82 Reichsstadt* – Anton Rast's fight book, although in this case Mair claims some editorial influence.

arms is adorned with a whip, which was rather crudely painted over with an anchor by another artist—possibly to look similar to the coat of arms of Paulus Kal (see for instance Bologna, *Ms. 1825*, fol. 5ᵛ; Gotha, *Ms. Chart. B. 1021*, fol. 1ᵛ)—for an unknown reason. After all, the codex differs considerably from Talhoffer's other manuscripts (K, B, C, M) in many ways, as HANS-PETER HILS discovered more than 30 years ago.[249]

Nevertheless, being a compilation, the various fighting disciplines appear overarchingly in many independent sources. HILS provides us with a detailed concordance of the images in the manuscripts he examined (G, K, V, B, C, M).[250] His overview, however, is flawed insofar as some images show generic postures that indeed look similar but only to an extent that is to be expected in a discipline such as fighting with large duelling shields, in which certain positions are commonplace. Upon closer examination, the similarities in this sequence are only superficial or coincidental. This is also true for harness fencing, see for instance fol. 56ʳ in G (see Fig. 14) and fol. 5ᵛ in the Cracow *Gladiatoria* manuscript (see Fig. 13).[251]

Figure 13 (above): Preparing to parry a thrown spear. Legend: Gladiatoria, 1440s. Crakow, Biblioteka Jagiellońska, Ms. Germ. quart. 16, fol. 5ᵛ.

Nevertheless, occasionally there is a striking resemblance among Talhoffer-related manuscripts. This is particularly noteworthy with a technique called "the scissors" which is shown almost identically in G (see Fig. 7), C (see Fig. 10) and M (see Fig. 11), and in a mirrored version in B (see Fig. 9) and K (see Fig. 8). A similar but not identical version can be seen in *Gladiatoria* (fol. 22ᵛ), but here the kick to the backside is omitted. In summary, we cannot determine whether Talhoffer actually wrote G, but he certainly was the owner of the manuscript.

Thrusts of doubt and vowel strikes

It can be fun to make an attempt to reconstruct master Talhoffer's teachings. But mostly, it is a somewhat tedious effort, due to the utter lack of exhaustiveness of the sources. His short

---

[249] HILS (1983), p 102.
[250] HILS (1985), appendix.
[251] Cracow, *Ms. Germ. Quart. 16.*

Figure 145: The thrust of doubt (leftmost figure).
Legend: Talhoffer, 1467. Munich, Bayerische Staatsbibliothek, Cod. icon. 394a, fol. 40ʳ.

captions offer little insight into the required movements, and much guesswork ensues. If, however, one has begun to develop an understanding of medieval fencing arts by the means of manuscripts by more verbose masters, such as Liechtenauer for instance, one may indeed come to more or less plausible solutions. It is particularly advantageous if a certain technique like the *Krummhaw*, which is shown in Talhoffer's 1467 codex M, fol. 11ʳᵛ, is explained in detail by Liechtenauer's commentators. However, it is almost impossible to make sense of other techniques.

In order to pick two examples, let us briefly examine the *Zwyuelstich* (thrust of doubt) and the *funff focal höw* (five vowel strikes).

The thrust of doubt is a unique thing: its name appears only once in M on fol. 40ʳ. Although performed by a fighter sans armour, the context makes it very clear that it is a technique intended for armoured combat with the sword. The combatant holds his sword by the blade with both hands, the handle protruding over the right ear, the point directed at his adversary's breast or groin. The posture seems to be more in line what in other situations is called a *Mordschlag*—a "murder strike" with the pommel of the sword. This particular image is very difficult to interpret, to say the least. No further text offers us any insight, nor can we gather any from other related or even unrelated sources. For the time being, this technique remains a complete mystery.

The other example refers to five vowel strikes. Unlike the previous example from M, this is a text passage from C and doesn't have an accompanying image. The text on fol. 2r says: *Der höw sind fünff und haissent funff focal, Die lern recht und mercks fürwär.* — "There are five strikes, and they are called five vowels. Learn them correctly and truly remember them." This is not an easy passage to grasp. On the one hand, the two lines don't rhyme, unlike the rest of the text in this section; on the other, interpreting *focal* as 'vowel' (*Vokal* in modern German) may be a bit daring at first glance. But then again, taking late medieval German orthography into consideration, maybe it's not: the letter 'v' was used either as a vowel or as a consonant—as it used to be in Latin. In German, the vowel 'v' was pronounced 'u', the consonant as either 'v' or 'f'. So we see that it's actually not too far a stretch to come to that conclusion.

As far as the actual strikes with the sword are concerned, master Liechtenauer has the "five hidden strikes" (which were re-phrased as five "master strikes" a century later): *zornhaw, krumphaw, twerhaw, schilhaw, schaitelhaw* (in order of appearance, quoted according to *Cod. 44 A 8*). Orthography was, as mentioned above, sometimes arbitrary, and particularly the final strike appears frequently as *scheittelhaw*, but if we stick with the example given, we notice that every strike contains one distinct vowel: *zOrnhaw, krUmphaw, twErhaw, schIlhaw, schAitelhaw*. Furthermore, if we bring the vowels in their correct order, we will find, that an astonishing sequence ensues: **A**, the *schaitelhaw* has hands and point of the sword in the highest position; **E**, the *twerhaw* still has the hands high, but the point can also attack low targets; **I**, the *schilhaw* has the hands slightly lower, with the point tilted downwards; **O**, the *zornhaw* has the hands in a middle position, point slightly upwards; **U**, the *krumphaw* finally has the hands in a middle position as well, with the point going downwards. So when we follow Liechtenauer's five hidden strikes in alphabetical order, we come from the highest to the lowest position.

I have no idea whether this is pure coincidence, whether Liechtenauer was already aware of this circumstance, whether Talhoffer made this up, or whether it is only me, as a practitioner, interpreting too much into the old verses. It is a fun thought, though, and a great way of providing modern students of the old art with a handy mnemonic. Interesting to see, by the way, that this bold idea of mine, spread on the internet a couple of years ago, has already been echoed in genuinely serious dissertations.[252]

In any case, Talhoffer's manuscripts still hold many a secret that is worthwhile to examine and explore. The more

---

[252] LEISKE (2018), p. 148.

information we can gather about the fencing systems of the old master, the more likely we will come to solid and plausible solutions.

Bibliography

Manuscripts

Augsburg, Universitätsbibliothek, *Cod. I.6.2°.1*
Berlin, Kuperstichkabinett der Stiftung Preußischer
    Kulturbesitz, *78 A 15*
Bologna, Biblioteca Universitaria, *Ms. 1825*
Coburg, Kunstsammlungen der Veste Coburg, *Hz.014*
Copenhagen, Det Kongelige Bibliotek, *Thott 290 2°*
Cracow, Biblioteka Jagiellonska, *Ms. Germ. Quart. 16*
Dresden, Sächsische Landesbibliothek, *Mscr. Dresd. C 487*
Glasgow, R. L. Scott Collection, *E.1939.65.341*
Gotha, Forschungsbibliothek Schloss Friedenstein, *Ms. Chart.*
    A558
Gotha, Forschungsbibliothek Schloss Friedenstein, *Ms. Chart.*
    B1021
Göttingen, Niedersächische Staats- und Universitätsbibliothek,
    2° Cod. Ms. philos. 61
Kassel, Landesbibliothek, *2° Ms. iurid. 29*
Königseggwald, Gräfliches Schloss, *Hs. XIX, 17-3*
Leeds, Royal Armouries Museum, *I.33*
Munich, Bayerische Staatsbibliothek, *Cgm 1507*
Munich, Bayerische Staatsbibliothek, *Cod. icon 394a*
Munich, Bayerische Staatsbibliothek, *Cod. icon 394*
Munich, Bayerische Staatsbibliothek, *Cod. icon 395*
Nuremberg, Germanisches Nationalmuseum, *Cod. Hs. 3227a*
New York, Metropolitan Museum of Art, *26.236*
Rome, Biblioteca dell'Accademia Nazionale dei Lincei e
    Corsiniana, *44A8*
Vienna, Kunsthistorisches Museum, *KK 5342*
Wolfenbüttel, Herzog August-Bibliothek, *Cod. Guelf. 125.16 Ex-*
    trav
Private collection, olim: Vienna, Österreichische Nationalbib-
    liothek, *Cod. Vindob. Ser. Nov. 2978*

Additional sources

BURKART (2014)
    ERIC BURKART: 'Die Aufzeichnung des Nicht-Sagbaren.
    Annäherung an die kommunikative Funktion der
    Bilder in den Fechtbüchern des Hans Talhofer', in
    JASER, CHRISTIAN/ISRAEL, UWE (eds): *Zweikämpfer.*
    Fechtmeister–Kämpen–Samurai (Das Mittelalter.
    Perspektiven mediävistischer Forschung, Band 19, Heft
    2), De Gruyter, Berlin, 2014

HAGEDORN (2016)
> DIERK HAGEDORN: 'German Fechtbücher from the Middle Ages to the Renaissance', in DANIEL JAQUET/KARIN VERELST/TIMOTHY DAWSON (eds): *Late Medieval and Early Modern Fight Books*, Brill, Leiden/Boston, 2016

HILS (1983)
> HANS-PETER HILS: 'Die Handschriften des oberdeutschen Fechtmeisters Hans Talhoffer', in *Codices manuscripti*, Hollinek, Vienna 1983

HILS (1985)
> HANS-PETER HILS: *Meister Johann Liechtenauers Kunst des langen Schwertes*, Peter Lang, Frankfurt am Main/Bern/New York, 1985

ISRAEL (2017)
> UWE ISRAEL: 'Die Fechtbücher Hans Talhofers und die Praxis des gerichtlichen Zweikampfs', in ELISABTH VAVRA/MATTHIAS JOHANNES BAUER (eds): *Die Kunst des Fechtens*, Universitätsverlag Winter, Heidelberg, 2017

LEISKE (2018)
> PATRICK LEISKE: *Höfisches Spiel und tödlicher Ernst—Das Bloßfechten mit dem langen Schwert in den deutschsprachigen Fechtbüchern des späten Mittelalters und der frühen Neuzeit*, Thorbecke, Ostfildern, 2018.

LENG (2008)
> RAINER LENG, HELLA FRÜHMORGEN-VOSS, NORBERT H. OTT, ULRIKE BODEMANN, PETER SCHMIDT, CHRISTINE STÖLLINGER-LÖSER: *Katalog der deutschsprachigen illustrierten Handschriften des Mittelalters* Band 4/2, Lfg. 1/2: 38. Fecht- und Ringbücher, C. H. Beck, Munich 2008

MÜLLER (1992)
> JAN-DIRK MÜLLER: 'Bild—Vers—Prosakommentar am Beispiel von Fechtbüchern. Probleme der Verschriftlichung einer schriftlosen Praxis', in HAGEN KELLER, KLAUS GRUBMÜLLER, NIKOLAUS STAUBACH (Eds): *Pragmatische Schriftlichkeit im Mittelalter*. Wilhelm Fink, Munich, 1992

PAURENFEINDT (1516)
> ANDRE PAURENFEINDT: *Ergrundung Ritterlicher kunst der Fechterey*, Hieronymus Vietor, Vienna, 1516

STANGIER (2009)
> THOMAS STANGIER: 'Ich hab hertz als ein leb ...—Zweikampfrealität und Tugendideal in den Fechtbüchern Hans Talhoffers und Paulus Kals', in *Ritterwelten im Spätmittelalter—Höfisch-ritterliche Kultur der Reichen Herzöge von Bayern-Landshut*. Museen der Stadt Landshut, 2009

WIERSCHIN (1965)
> MARTIN WIERSCHIN *Meister Johann Liechtenauers Kunst Des Fechtens*. C. H. Beck, Munich, 1965

Online sources (accessed 23 March 2018)
  wiktenauer.com/wiki/Hans_Talhoffer
  talhoffer.wordpress.com/category/a-life-like-that-of-talhoffer/
  biography/

3. Six Weeks to Prepare for Combat: Instruction and practices from the Fight Books at the end of the Middle Ages, a note on ritualised single combats

Daniel Jaquet

This contribution is following the first axis of the argumentation of the conference *Shaping Bodies for Battle.*[253] Several historians of medieval warfare complained about the lack of sources to delineate the training of men for battle,[254] as opposed to later periods, where a dedicated body of literature is available for the study of the soldiers' training.[255] However, in the specific context of late medieval ritualised single combat, including the various phenomena labelled as "judicial combat", some relevant information can be analysed to address the questions related to the preparation for battle, not only in narrative, but also in technical sources.

These types of combat were indeed considered as "battle" (a dedicated form of warfare) in the late Medieval mindset,[256] as they dealt with serious matters (*Ernst*) contrary to other forms of ritualised combat in a more playful fashion (*Schimpf*),[257] such as knightly games (tournaments, *pas d'armes*, jousts, etc.), or urban feasts including competitive sporting praxes (fencing schools, wrestling matches, mock battles etc.). However, as argued in this contribution, all of those types of combats need to be studied together as part of the same phenomenon, diffracted by cultural habitus and societal contexts.

Therefore, I propose a study of selected instruction for the preparation for "judicial" combat found in the heterogeneous corpus of the fight books in the late Middle Ages,[258] which discuss not only knightly or princely

Figure 1: Mounted duelists Legend: Anonymous, 1437. Vienna, Österreichische Nationalbibliothek, Cod. 3062, fol. 27r.

---

* I am thankful to Ariella Elema for her comments and corrections.

[253] Underlying questions were formulated as follows: Which practices were used to make bodies fit for battle? What bodily techniques were taught and trained? What was seen as a fighter's ideal bodily appearance? How did fighters physically experience the shaping of their bodies?

[254] For example, CONTAMINE, 2003, p. 364; KEEN, 1984, p. 238. See the recent contributions to this question in the very late 15thc and early 16th c. in DERUELLE/GAINOT, 2013.

[255] For a study of technical literature on the training of the soldier, see LAWRENCE, 2009.

[256] For example, judicial combats are considered as such in the treatises of John of Legnano (1320-1383), Honorat Bovet (1340-1410), and Paris de Puteo (1410-1493). For a general discussion of these authors, see CAVINA, 2005. pp. 41-106.

[257] For a comparison of those two types of circumstances in the perspective of ritualised combat, see JAQUET, 2016b and JASER/ISRAEL, 2014.

[258] See the contribution of E. Burkart in this volume, as well as JAQUET, 2016a; BOFFA, 2014 and ANGLO, 2000.

single combats, but also those of the lower social strata. Moreover, I shall highlight three examples of those single combats for commoners out of a selected corpus of narrative, normative and pictorial documents in different geographical locations in the 15th c., in order to illustrate the instruction for combat found in the Fight Books. Both the relation of those fights and the technical content of the Fight Books also provide information about the fighting itself, which is less interesting for the purpose of this contribution. As a concluding point, the mutilation of the fighters as one common feature of the narration of the cases chosen can be highlighted in order to address issues concerning the representation of ritualised violence and about bodies in battle relevant for other interests in this volume.

Fighting for the truth, the proof of his cause or his honour

Figure 2: Duel between Jean de Carrouges and Jacques le Gris. Legend: Jehan de Wavrin, Anciennes et nouvelles chroniques d'Angleterre, 1470-1480. London, British Library, Royal MS 14 E IV, fol. 27r.

A widely accepted scholarly trend points out that the judicial duel or trial by battle tended to subside during the 15th c. to be replaced by the later duel of honour.[259] Recent works demonstrating that these forms of ritualised combat did indeed evolve, but with less rupture than was earlier believed have refined this postulate.[260] The point has been made that both temporal and spiritual powers attempted to limit the trial by combat as a lawful way of resolving conflicts, alongside the prohibition of the ordeal from the end of the 12th c. onward.[261] However, the aristocracy who still claimed to be entitled to resolve conflicts through ritualised combat as part of their hereditary and status privileges raised strong opposing voices.[262] The thirteenth and fourteenth centuries witnessed repeated prohibitions or limitations as authorities attempted to constrain those types of combat to very specific cases.[263] The practice of hiring a champion to settle legal grudges also appeared to decrease.[264] The late fourteenth century combats of Carrouges versus Le Gris[265] (see Fig 2.) or Estra-

[259] NOTTARP, 1956; MOREL, 1984.

[260] CAVINA, 2003 and 2005; ISRAEL, 2008; HILTMAN/ISRAEL, 2007; ISRAEL/ORTALLI, 2009; NEUMANN, 2010; ELEMA, 2012; LUDWIG et al. (eds.), 2012.

[261] MCAULEY, 2006, pp. 473-513.

[262] STANGIER, 2009, pp. 75 and 78.

[263] See the discussion about those circumventions in HILTMAN/ISRAEL, 2007, pp. 65-84 and TELLIER, 2012, pp. 107-121.

[264] See the discussion about the historiography in JASER, 2014, pp. 380-406. See also ISRAEL, 2008, pp. 121-147.

[265] GUENÉE, 1995, pp. 331-343 and PARAVICINI, 2016, pp. 23-84.

vayer versus Grandson[266] are believed to be among the "last" trials by combat for the aristocracy and, if the often quoted passage of Olivier de la Marche[267] is to be trusted, would have fallen into disgrace by the end of the fifteenth century. However, "judicial combat" was still found in the 15th c. as a procedure codified in various customary laws, as an object of discourse in the heralds' compendia[268] and a direct application of the martial gesture in technical literature.

This apparent paradox is due to the problematic definition of those types of combat and the focus of scholarly research on the high strata of society. Indeed, the involvement of high aristocracy in the "old" trial by combat was exceptional during the fifteenth century,[269] even if it was still an object of discourse and provocation among princes.[270] Various customary laws had codified procedures for judicial single combat (lat: *duellum*, mfr: *gage de bataille*, mhd: *kampf*) for all social strata since the 13th c. (but mostly in the 14th c.),[271] mainly dealing with legal matters and procedures before the combat itself. These *consuetudines* also led to specialised *Kampfrecht*, especially in franchised cities during the second half of the 14th c., throughout the 15th c. and up to the middle of the 16th c.,[272] where details of the ritual before and during the combat itself were codified, including weaponry, clothing, the role and function of the different actors involved, etc. These single combats were settled by judicial courts, according to customary law and were fought on foot with long shields and either a wooden mace or a sword.[273] For example, the custom of Zwickau mid-14th c., drawing from

---

[266] BERGUERAND, 2008.

[267] [...] *car peu de gens vivans ont veu l'exécution de gaige de bataille, et a plus de soixante et dix an que, soubz ceste maison de Bourgongne, ne fut telle œuvre exécutée entre deux nobles hommes. Et moy qui ay demouré en ceste noble maison près de soixante ans, je ne veis de ma vie gaige de bataille.* Ed. PROST, 1878, p. 20. To be noted that this quote is only valid for aristocracy (entre deux nobles hommes). La Marche witnessed at least one judicial combat between commoners (Plouvier and Coquel, discussed below).

[268] As defined by HILTMANN, 2011. See also his article with U. Israel on the specific correlation between those types of combat and his corpus of sources (HILTMANN/ISRAEL, 2007, pp. 65-84).

[269] For example the case of the Earl of Ormond versus the Prior of Kilmayne in London in 1446 or Hector de Flavy versus Maillotin de Bours in Sedan in 1430 (see ELEMA, 2012, p. 161, note 79 and p. 312). See also the examples in the Holy Empire quoted from the dissertation of POSCHKO in JEZLER, 2014, p. 188.

[270] See examples in VONES, 1996, pp. 321-332.

[271] The best known examples among legal historian are Eike von Repgow, Sachsenspiegel (ed. ECKHARDT, 1972, Landgericht I, 63), or the Rechtsbuch Kaiser Ludwig von Bayern, 1346 (ed. VOLKERT 2010). For the French kingdom, Philippe de Beaumanoir, Coutume du Beauvaisis. Also found as specialised treatises in the Italian peninsula, such as the anonymous *Summula de Pugna* or Roffredus of Benevento, *Summa de pugna*. This is however not exhaustive and there is no reference study on those consuetudines for the judicial combat taking in account all those geographical areas. For the Holy Roman Empire, see NEUMANN, 2010; for the French kingdom, see TELLIEZ, 2011 and CARBASSE, 1975, pp. 385-403; for Italy, see CAVINA, 2003.

[272] See example quoted in FORTNER, 2007, pp. 10-22. For a more comprehensive discussion of the *Kampfrecht* in the South Germany, see LEISER, 1986, pp. 5-17. However, this phenomenon is not limited to the Holy Roman Empire, but stems to the kingdoms of England, France, Spain and many cities in the northern Italian peninsula.

[273] In England and parts of France, another kind of weapon seems to have been customary: the *baculus cornutus*. See ELEMA, 2012, p. 249.

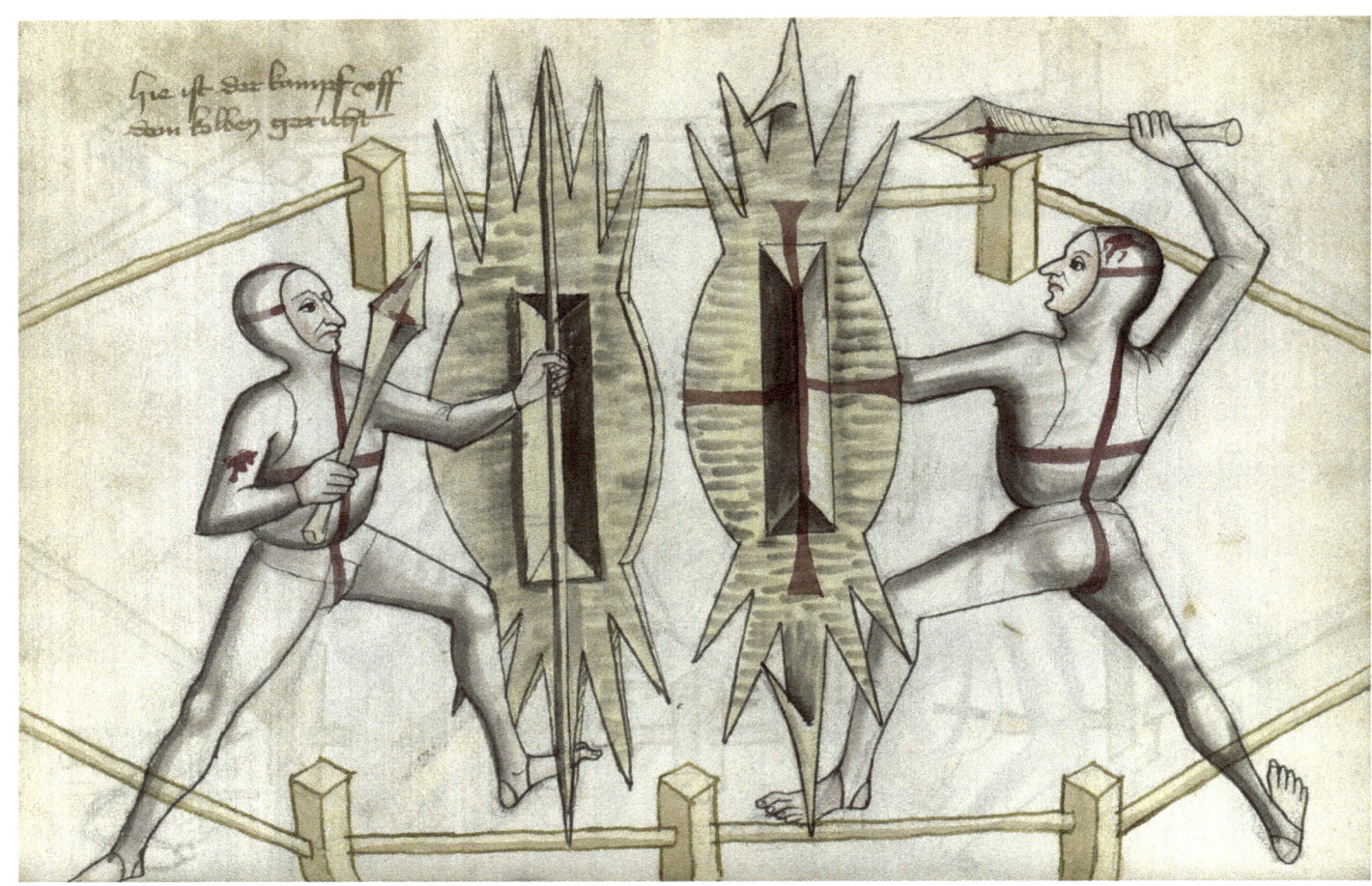

Figure 3: Franconian custom. Legend: Hie ist der kampf uff dem kolben gericht. Kobenhavn, Det Kongelige Bibliotek, Thott 290 2°, fol. 99v.

material out of the *Sachsenspiegel*, advises that "All knights, valets (*knecht*) and merchants shall fight with a sword", while "Peasants shall fight with a wooden mace".[274]

Iconographical sequences depict such judicial single combats with the associated technical repertoire in several fight books of the 15th c.[275] For the purpose of this article, I shall focus on those attributed to Hans Talhoffer and Paulus Kal, two 15th c. fencing masters (*Schirm-, Fechtmeister*).[276] The preparation for judicial combat appears to have been part of their professional trade, according to the content of some of their written productions dedicated to the low and high aristocracy of the South Rhinelands.[277] Of particular interest are

---

[274] *Alle rittere, knechte und kauflüte sullen vechten mit dem swerte. / Alle gebüre sullen vechten mit kolben.* Zwickauer Rechtbuch 1348-1358, ed. PLANITZ, 1941, vol. II, 26, 6-8.

[275] To be found in fight books attributed to Peter Falkner (Wien, Kunsthistorisches Museum, KK5012), Jorg Wilhalm (Augsburg, Universitätsbibliothek, Cod.I.6.2°.3 and Cod.I.6.2°.2; München, Bayerische Staatsbibliothek, Cgm 3711 and 3712), Paulus Hector Mair (Dresden, Sächsische Landesbibliothek, Hs Dresd. C93/94; München, Bayerische Staatsbibliothek, Cod. Icon.393 1/2; Wien, Österreichische Nationalbibliothek, Wien, Cod. 10825/ 10826) and in anonymous compendia (Paris, Musée National du Moyen Âge, Cl. 23842; Wolfenbüttel, Herzog August-Bibliothek, Cod. Guelf. 78.2 Aug. 2°). For a focus on the judicial combat between man and woman, I am preparing an article on the matter for the journal "Le Moyen Age" (forthcoming).

[276] T. Stangier presents them as rivals, although there is no evidence that they were ever in contact. See STANGIER, 2009, pp. 79-93. For Talhoffer, see the historiographical review in BURKART, 2014, pp. 253-301. For Kal, see for instance WELLE, 1993, pp. 240-255.

[277] For a discussion of the dedicatees, see STANGIER, 2009, pp. 79-93 and BURKART, 2014, pp. 253-301.

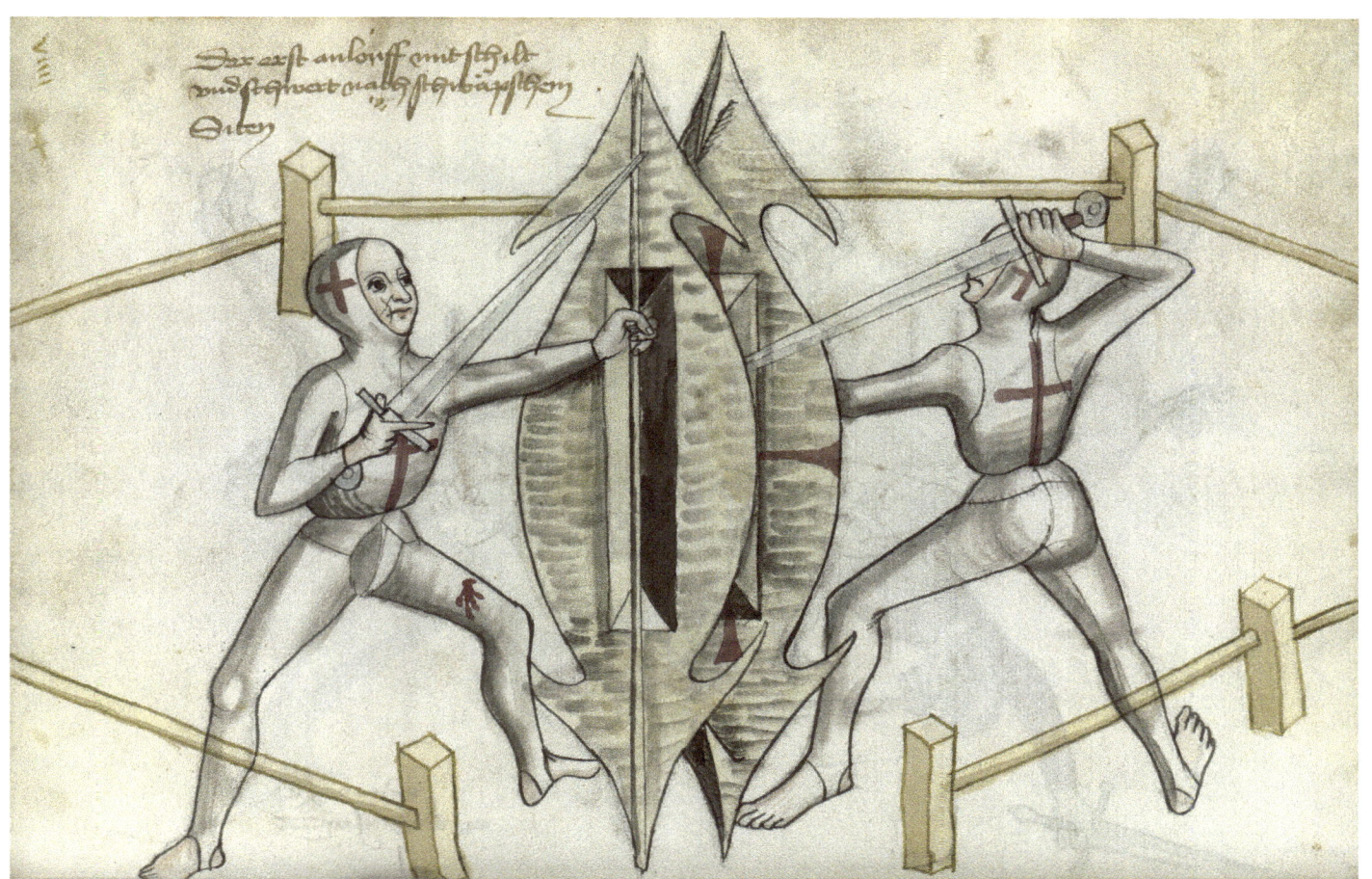

*Figure 4: Swabian Custom.
Legend: Der erst anlouff mit
schilt und schwert nach
schwäpschen Siten. Koben-
havn, Det Kongelige Bibliotek,
Thott 290 2°, fol. 97r.*

several passages compiled in some of their compendia,[278] edited
in appendix (1, A-C) and discussed below. Both masters dealt
with knightly judicial combat (in armour on foot), judicial combat for commoners or burghers (including combat for a man
against a woman). I shall concentrate on the second category of
combat, which distinguishes between two technical repertoires:
one involving a shield and sword—Swabian custom, the other a
shield and mace—Franconian custom (see Fig. 3-4).[279]

The training of clients for "judicial" combat as a specific kind of trade for fencing masters

Both masters were well versed in the institutional underpin-
nings of those types of combat. Hans Talhoffer underlines that
they still occurred within cities according to customary laws,
although the *Decretals*[280] forbade those combats and emperors,

---

[278] References in appendix. Apart for those quoted, a large manuscript tradition is attributed to the authors,
for description see LENG, 2008, 38.3 (Talhoffer) and 38.5 (Kal). There is however one manuscript misattributed
to Talhoffer (38.3.7, which is in fact a copy of Kal) and the list of the copies is not exhaustive.

[279] For a short and incomplete description of both customs, see FORTNER, 2007, pp. 19-22. See also some references in STANGIER, 2009, p. 75.

[280] Those are the Decretals of Raymond of Peñafort. After Pope Innocent III and the Fourth Lateran Council of
1215 prohibited clerics from attending or taking part in judicial duels, (*Constitutiones quarti Lateranensis una
cum commentariis glossatorum*, ed. GARCIA, 1981, C.18), Pope Gregory IX issued Peñafort's Decretals in 1234
(IBID., X 3.50.9).

Figure 5: Illustrations of preparations for dueling.
Legend: Talhoffer, 1448. Gotha, Forschungsbibliothek Erfurt/Gotha, MS Chart.A.558, fol. 35v, 37v, 38v-39r, 40r.

princes and lords frowned upon them (A1). Stating that such endeavours are affairs of honour (*mütwill*),[281] he then lists the "legal" reasons to undertake such single combats: murder, treason, heresy, acts of betrayal towards one's lord, betrayal of one's given word when captured, fraud and the abuse of a woman (A2). As Jezler noted, these causes are comparable to those forbidden and punished by contemporaneous tournament societies, as a means to maintain the noble ethos through social regulation.[282] This fact also highlights the similarity between the different forms of ritualised combat, of both serious and more playful varieties, touching upon different social strata.

Hans Talhoffer has several detailed insights for his client regarding the procedures of the ritual, before the combat (A3-A5)—including how to act in cases which may lead to the cancellation of the fight (A6-A7)—and at the start of the combat (A8-A9, A15-A16). Several of those procedures are also depicted in different versions of the fight books (see Fig. 5 for one example).[283]

As a professional in this field, Paulus Kal also compiled a list of items to be clarified with the court (including the judge, *urtailer*) by the master or his client in preparation for the combat (C1-26). This list is reminiscent of the so-called *Questions,* which Geoffroy de Charny addressed to his peers, the knights of the Order of the Star, one century earlier and which were related to procedural questions concerning the law of arms and practice in tourneying or jousting.[284] In Charny's case, the researcher would love to have the matching answers, but the analysis of such questions allows one at least to outline several praxeological elements by deductive analysis.

In terms of Kal's questions, the different actors and their roles during the combat can be deduced (C1) as follows: the adviser (*warner,* C2-3) interacts with court officials on behalf of the combatant; the listener (*lüsner,* C5, 8) is an assessor at court witnessing the combat; and the grid-warden (*grieswartl* C8-10) is the second of the combatant, allowed within the barriers and equipped with a staff.

---

[281] See the discussion of this matter in JAQUET, 2016b.

[282] JEZLER, 2014, p. 189, ref. to pp. 66f.

[283] Several iconographical cycles depicting the rituals before, during and after the combat, usually in the sections dedicated to the armoured combat on foot with swords and unarmoured combat on foot with judicial shield and sword or mace. For a description of the different versions of the works attributed to Hans Talhoffer, see BURKART, 2014, pp. 253-301.

[284] See KAUEPER/KENNEDY, 1996.

The combatant has the right to call his grid-warden to "pull the staff", so that the fighting is interrupted and each party can go to their rest-place (C7/C18). There is a respective example in the *Kampfrecht* of the city of Gelnhausen from 1360, where the combatant can call for the staffs three times.[285] Other details of relevance are, for example, how many maces the combatant can have (C13), what happens when he loses them inside or outside the circle, who is allowed to hand them back to him and when they are allowed to do so (C14-16). The conditions of defeat also require examination in detail (C17, 18, 24). For example, what is considered stepping out of or being pushed outside of the circle: "is it a hand, the body, a foot, the mace or the shield"? (C17). Final questions address issues about the crowd attending the fight, what regulation is there about the need to remain silent during the fight (C25) and what are the precautions taken to isolate the combatant from the crowd (C26)? Again, this is very similar to the regulations for tournaments or chivalric games, including single combat, which all attempt to regulate the reaction of the crowd in attendance.[286]

Concerning specialised weaponry and clothing, Kal and Talhoffer provide relevant information not only in the iconographic sequences depicting fighting techniques, but also in the text. Kal suggests inquiring about this subject before the fight (C19, C20). The Talhoffer compendium of 1459 comprises detailed (technical?) depictions of the clothing itself (107[r]), the maces (106[v]) and five different shields (104[r]-105[v]) with short written comments. Talhoffer also describes how the combatant should enter the barriers, with details on the clothing (A16) and the gesture performed for different parts of the rituals taking place prior to the fight.[287] Complementary details describing these outfits (discussed below) can be found in narratives and some normative texts of customary law.[288]

---

[285] *So bit er fragen, wie dicke er den stangen begeren solle, so wird erteilt: dry stunt , diewyle sie sich nit begrifen haben, wan aber sie sich begryfen, so mag er ir keiner me begeren.* Hessisches Urkundenbuch, ed. HIRZEL, 1894. Credit for this finding goes to Jens-Peter Kleinau who published a blog article about it in 2013.

[286] JEZLER, 2014, pp. 57-72. For a comparison between different rulesets, see RÜHL, 2001, pp. 193-208.

[287] Some of those rituals are also described in heraldic compendia. See ISRAEL/HILTMANN, 2007, pp. 65-84. Also of interest for anthropological studies of those rituals, outlined in the perspective of an historian, is OSCHEMA, 2011, pp. 142-161.

[288] For example in 1446: *Alß dem scheffen ingeben yst, wie die zu dem kamp geschickt und gestalt sin sullen sin, hat der schieffe gewisset, sie sullen haben zum ersten eyn graen fyltzrocke, kogeln, hossen und schue; an eym ander zwen glych schilde, iglicher ein holczen kyckel und eyn hanthabe und sullen in wynden under dem kyne, zwen holtzen kepel glich eln lang, dryeecket hynden yen knop, und zwen henen degen glich eyner halben eln lang in eym fure und den spitzen gehert und die hencken by den uff die rechten sytten zu der hant um.* Landschreibereirechnung der Obergrafschaft (Landgericht Katzenelnbogen), DEMANDT, 1953, vol. 3, p. 2294. I thank Christine Reinle for having pointed out this source to me.

Figure 6: Physical training. Legend: Talhoffer, 1448. Gotha, Forschungsbibliothek Erfurt/Gotha, MS Chart.A.558, fol. 29ᵛ.

Physical training, diet of the combatant, and the need for secrecy

According to Talhoffer, when a complaint has been lodged at the court in the appropriate manner (A3), "six weeks will be granted for his training days, and also four more days before his judgement, so that they can fight according to the custom of the land and the law" (A4). The hiring of fencing masters and the training period of six weeks are also found in customary law, for example in Münstemaifeld in 1372.[289] According to Talhoffer, both parties are bound by custom and the court not to break peace during this period under penalty of banishment (A4).

In the first compendium attributed to the master (1448), on the versos of the folios showing the technical sequence of the fighting gestures, an iconographical cycle illustrates the entire process of preparation for a judicial duel (see Figs. 5 and 7), from the hiring of the master outside of the city walls, to the preparation of the fighter up until the combat. This included such distractions as eating, listening to music, bathing, hunting, and time spent with his relatives, but also ritualised processes such as shaving, praying and anointing (App. 2).[290] One of the illustrations in the cycle depicts two physical exercises, which may have been part of the training: stone and javelin throwing (see Fig. 6). These exercises are also found in another of his fight books, dedicated to Luitold von Königsegg, which contains general rhymed advice given to one who cares about the "knightly values": he should "train during peace time by throwing stones and javelins, dancing and jumping, fencing and wrestling, mock jousting and tourneying, and courting beautiful ladies".[291] The same book provides very rare information on physical training and diet in the context of preparations for a judicial combat:

---

[289] *halden sal vnvirderfflich ses wochen und dry tage, und yeme eynen meister gewinnen, der en kempen lere, und sal en halden, und daz alles dun der greue uff sine kost selber, ob der ghene der kuste nit enhat, der kemplich wirt angesprochen.* Quoted in NEUMANN, 2010, p. 90, note 414. Other quotes regarding the hiring of fencing masters (*Bretons*) according to Norman customs in COULIN, 1906, p. 87. For English cases, see RUSSELLL, 1984, pp. 76-78.

[290] See also STANGIER, 2009, p. 74.

[291] [...] *vnd gedenck nach ritterschafft / mit freiden ueben / stein werffen vnd stang schueben / tantzen vnd springen / fechtten vnd ryngen / stechen vnd turnieren / schónen frawen hofieren.* Ed. SCHULZE, 2010, p. 23 (revised by the author).

"He should especially get up early every day and hear a mass, then go back to his house, eat a loaf of St-John bread[292] and train for two hours. He shall not eat too much greasy food. After noon, he shall train two hours and at nightfall, before he lies down to sleep, he shall eat a slice of rye bread soaked in water. This makes for good breathing and a strong heart." (B2).

The master also states that he will evaluate his client to establish "whether he is weak or strong, choleric or gentle-minded, whether or not he has good breathing, and if he would work heartily" (A13). He should also know "how his top heats up if someone would quarrel or fight" (A14). On a side note, similar but more detailed advice based on humour theory and related to martial training is depicted in the fight book of Pietro Monte, written in the last decade of the 15th c., but published post-mortem in 1509.[293]

Talhoffer also notes that the client shall recognise the master as being trustworthy by the following qualities: pious, sober, righteous and protective, and his ability in the art of combat—"broadening the arsenal of techniques and knowledge about the art" (A12). Securing his trade, he warns against untrustworthy rivals, such as other fencing masters before him (and after him...).[294] He also insists on the need for secrecy: "Yet the combatant and the master shall guard that they let no man see them or the arsenal with which they work. And they both shall guard their doings from much of society; and say little of the fighting, so that no notice is made thereof." (A13)

This is, of course, crucial to the training for a judicial combat, but this kind of face-to-face instruction and need for secrecy is also emphasised by Fiore dei Liberi in other contexts. In his treatise from the very beginning of the 15th c., this master lists his students (among them knights) and explains that he trained them for deeds of arms (and for more serious matters). He then states that he has been "well paid" and that "he always taught this art in sec-

Figure 7: Diet.
Legend: Talhoffer, 1443. Gotha, Forschungsbibliothek Erfurt/Gotha, MS Chart.A.558, fol. 31ʳ.

---

[292] Bread baked for the feast of St John (December 27). St John's wine and bread are common feature of medieval recipes (for St John blessed a poisonous glass of wine to render it harmless).

[293] For an introduction to this text, see FORGENG, 2014, pp. 107-114. The master also wrote a treatise on the "distinction of men", with long development on the physiological attributes related to physical exercise. See FONTAINE, 1991, p. 46.

[294] A12. This kind of warning is found for example in the first witness of the *Zedel* of Liechtenauer at the end of the 14th c.; *Als man noch manche leychmeistere vindet dy do sprechen / das sy selber newe kunst vinden vnd irdenken vnd meynen das sich dy kunst des fechtens von tage czu tage besser vnd mere.* Ed. ZABINSKI, 2010, p. 130 (revised by the author).

recy".[295] The monetary value of those secret teachings was also regulated in urban context within fencing guilds.[296]

Gouge the eye out: St William's miracle

Two of these judicial combats between commoners are described in chronicles of the 15th c. One in Valencienne (duchy of Hainault) in 1455 opposed Jacotin Plouvier, a burger of the town and Mahiot Coquel, a tailor from Tournai; another in London in 1456 opposed James Fisher, a tailor and fisherman, with Thomas Whitehorne, a man with no known profession and with reputation for robbery.[297] Both deadly combats are related in very crude terms and the chroniclers all made clear that they disliked this type of "improper" single combats.[298] The chroniclers give interesting details regarding the clothing and the weapons, supplementing information found in fight books and normative texts. For example, the combatants, whose hair and nails were cut, wore a tightly fitted garment made of leather[299], covered in grease "so that they could not grapple each other" and their hands were covered with ashes "so that they could handle their shield and mace".[300] The maces are all described as made of hardwood (medlar for d'Escouchy—*mellier bien nouteilleux* and de la Marche—*mesplier*; newly cut ash for Gregory—*grene hasche*).[301] None of the

---

[295] [...] *di questi e d'altri i quali io fior ò magistradi io son molto contento perché io son stado ben rimunerato e ò aibudo l'onore e l'armore di miei scolari e di parenti loro digo anchora che questa arte io l'ò monstrado sempre ocultamente si che non glie sta presente alchuno.* Fiore de'i Liberi, Flos Duellatorum, 1409. RUBBOLI/CESARI, 2003, p. 25.

[296] For example in the Statutes of the Fencing Masters of Bruges in 1456: *Ende als van den verboorghene consten, te wetene ghewaepent te cechtene met haecsen end andersins, dat elc meester ende provoost boven dien van elcken leerlinghe zal moghen nemen dies hem ghebueren zoude moghen.* GALAS, 2011, p. 148. For a recent study on fencings guilds in the Lowlands, see GEVAERT and VAN NOORT, 2016, pp. 221-242.

[297] Both case are studied in ELEMA, 2012, esp. pp. 1-5, 68, 135f., 234-237, 247-253, 266, 307-309, 313f., 318, 324-326. The sources for the first duel are *Chronique* de Mathieu d'Escouchy (DU FRESNE DE BEAUCOURT, 1863, vol. 2, p. 297-305); *Mémoire* d'Olivier de la Marche (PETITOT, 1825, vol. 2, p. 213-218); *Chronique* de Georges de Chastellain (KERVYN DE LETTENHOVE, 1864, vol. 3, p. 41-49); for the second one: *Gregory's Chronicle of London* (ed. GAIRDNER, 1876, pp. 199-202). For the first case, see also CAUCHIÈS, 1999, pp. 655-668 and LECUPPRE-DESJARDIN, 2016, pp. 181-197.

[298] For example: [...] *tenoit en la bataille* [...] *plus honte que honneur* [...]. Olivier de la Marche, ed. PETITOT, 1825, p. 407; or the London chronicler: *.hyt ys to schamfulle to reherse alle the condyscyons of thys foule conflycte.* Ed. GAIRDNER, 1876, p. 200.

[299] [...] *moste be clothyd alle in whyte schepys leter, bothe body, hedde, leggys, fete, face, handys, and alle.* IBID., p. 200. *Ilz avoient les testes raises, les piedz nuz, et les ongles coppez des mains et des piedz; et au regard du corps, des jambes et des bras, ilz estoient vestuz de cuyr bouilly, cousu estroictement sur leurs personnes.* Olivier de la Marche, ed. PETITOT, 1825, p. 404.

[300] [...] *deux bassins plains de gresse, dont les habillemens ... furent oingtz et engressez, affin que l'ung d'eulx ne peust prendre prinse sur l'autre.* [...] *deux bassins de cendres, pour oster la gresse de leurs mains, afin qu'ilz puissent mieulx tenir leurs escuz et leurs bastons.* IBID. p. 405.

[301] The more detailed description in given in the Gregory's chronicle: "[...] and that they should have in their hands 2 maces of freshly cut ash, with bark being upon it, of 3 feet in length, and at the end ought to be a cudgel of the same[wood], provided that the addition adds any length at all." ([...] *grene hasche, the barke beynge a pon of iij fote in lenghthe and at the ende a bat of the same govyn owte as longe as the more gevythe any gretenys*). Gregory's chronicle, ed. GAIRDNER, 1876, p. 200. I thank Daria Izdebska for her help in translating that excerpt. This appears to be specific kind of mace, not to be compared with the ones used in France or Germany. See RUSSELL, 1983, pp. 432-442.

chronicles are illustrated during the 15th c. (some illustrations are found in the 16th c.),[302] but the manuscript of a 15th c. chronicle from Brabant[303] illustrates a different duel, from 1236, depicting contemporaneous fashions for tightly fitted garments and maces (see fig. 8).

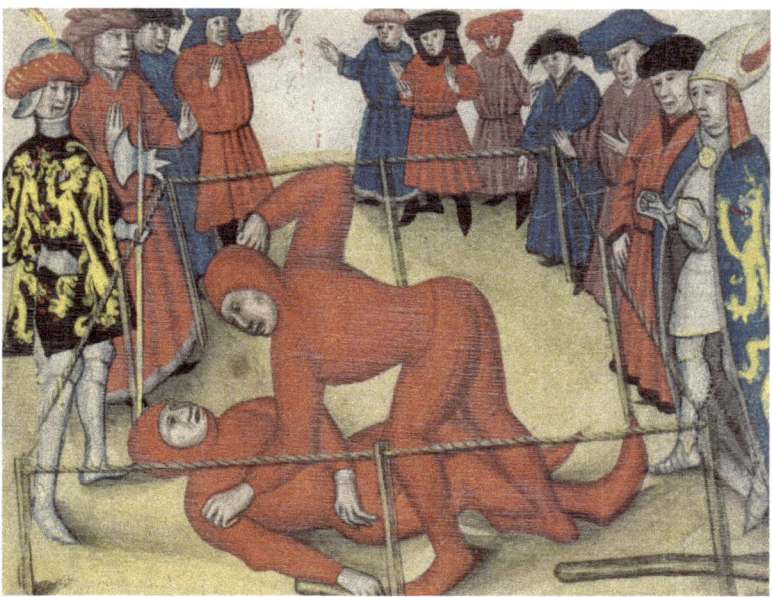

Figure 8: Judicial combat. Legend: Jan van Boendale, Brabantsche Yeesten, 1450-1480. Bruzelles, Biblioteque royale, Ms. IV 684, fol. 68.

The fight follows the same pattern in all these cases: both combatants exchange a few blows, then the fight proceeds to grappling, and it ends on the ground. All written accounts describe very crude moves, including biting, scratching, breaking limbs and eye gouging. In the Whytehorne-Fisher case, one of the combatants bit the other in the private parts before the other bit the first man's nose and gouged his eye.[304] In the Plouvier-Coquel case, when both men wrestled on the ground, one gouged out both his opponent's eyes, and crushed his stomach with his knees while strangling him to death.[305] These un-chivalrous gestures are mentioned in the fight books' repertoire, but are usually—when described or referred to—labelled as "forbidden" or "secret". In connection to these, various regulations of the fencing competitions (*fechtschulen*) list these prohibited fighting techniques in the late 15th and 16th c.[306] As the statutes of the fencing guild of Bruges demonstrate,[307] those gestures were taught. One can also assume that they were part of the client's training in preparation for judicial combat. As a side note in the Plouvier-Coquel case, the master attributed to Mahienot Coquel is

---

[302] One illustration of the Coquel and Plouvier found in a 16th c. manuscript by A. Elema (Douai, Bibliotheque municipale MS 1183, ff. 188v-189r). Also to be noted that Paulus Hector Mair, compiler of a large anthology of the art of fighting in the middle of the 16th c. did also include illustrations of such fights, as well as copies of *Kampfrechten* and precise drawings of the shields, see note 61.

[303] The redaction of this chronicle and the realisation of this manuscript constitute a complex case. The attribution to Jan van Boendale (Jan de Klerk) is dubious and one possible source for the relation of the duel is the chronicler Lodewijck van Vethem. The text has been edited by WILLEMS, 1837, pp. 26-32. I am grateful to Sergio Boffa for providing me with this information. He is currently preparing an article about this case.

[304] [...] *and bote hym by the membrys* [...] *and toke that fals peler by the nose with hys tethe and put hys thombe in hys yee. Gregory's Chronicle*, ed. GAIRDNER, 1876, p. 200.

[305] [...] *car de ses mains et ongles lui creva les deux yeux de la teste et fist saillir dehors. Aveux clui estant sur son estomach a genoux lui creva le cœur et l'estrangla de ses mains puis le jetta hors desdictes lices.* Escouchy, ed. DU FRESNE DE BEAUCOURT, 1863, p. 297. See also for the eye gouging: [...] *Sy alla bouter son pousse de l'autre main en ses yeux et y commença à fouiller dedens jusqu'au parfont tellement qu'il lui tirat les yeux hors jusques à pendre sur les joues.* Chastellain, ed. KERVYN DE LETTENHOVE, p. 48. An additional manuscript of the chronique of Chastellain describes at length in very crude details the end of the combat (London, British Library, Additional Ms 54156, ed. DELCLOS, pp. 325-327).

[306] For examples of those techniques and discussion of the connexion between fight books and fencing schools, see JAQUET, 2013.

[307] See footnote 44.

Figure 9: The eye gouging.
Legend: Stained glass panel
of Saint William, detail.
York Minster, 1414.

named "Hans" in the Chronique of Chastellain.[308] Of course, without other evidence, this is unlikely to be Hans Talhoffer.

Finally, an interesting common feature of the written descriptions of these combats is the eye gouging. A series of stained glass panels in York Minster's Saint William window also illustrate this gesture (See fig. 9).[309] This is related to one of the miracles in the *Vita* of Saint William of York (?-1154, canonised by Honorius III in 1227), here depicted as an early 15th c. judicial combat.[310] In this miracle, Ralph, falsely accused of breaking the king's peace, has to fight in a judicial combat. His eye is gouged out by his mightier opponent Besing. The sentence of the court is the loss of the other eye. The blind victim recovers his sight while visiting the tomb of Saint William. In addition to the *Vita*, the fact that all narrative accounts studied here include eye gouging may point towards a topos for judicial combat.[311] However, this remains a postulate to be investigated by further studies, outside of the scope of this contribution.

Conclusion

This contribution highlights some of the benefits of further investigation of this type of judicial single combat for commoners in order to address issues related to the training of combatants as well as bodies in battle. As outlined by the different cases mentioned, these praxes occurred across the Western Europe at the end of the Middle Ages.[312] Further research may reveal that these were still practiced in the 16th c.[313] and certainly were not limited to the geographical boundaries of this contribution. These understudied single combats are indeed part of a greater phenomenon of ritualised combat,

---

[308] *Or ont eu ces deux gens-icy par longue espace leurs maistres emprès eux, qui leur ont appris leurs envayes et deffenses, et tout ce en quoy il les espèrerent à sauver, et avoit Mahienot empès lui un nommé Hans, le meilleur qu'on savoit en nul pays,* [...] Chastelain, ed. KERVYN DE LETTENHOVE, pp. 44f..

[309] FRENCH, 1999, pp. 70-73. Credit for this finding goes to A. Elema who mentions it in a note in her dissertation, see ELEMA, 2012, p. 255, n. 67.

[310] Miracula 37, See NORTON, 2006, pp. 169-181 and 198-200.

[311] See the discussion about castration in combination with blinding as a Norman punishment in ELEMA, 2012, pp. 152-155.

[312] The different examples in this contribution occurred in England, France, Germany, and the Low Countries.

[313] In the context of the redaction of the fight books, the specialised sections about judicial combat and those about armoured combat on foot were still compiled up to 1570 (see JAQUET/WALCZACK, 2014). The fight books of Paulus Hector Mair (three versions between 1540-1550) would prove interesting, since they also contain various *Kampfrechte* and sections dedicated to judicial combats (see note 50).

regulated by various norms and cultural habitus, most of the latter being tacit or implicit. The various forms of chivalric games—as models or ideals—also affect other types of sporting, competitive or ludic praxes involving all strata of the society, from the rural, to urban, and even to courtly contexts. The need to revise historiographical misconceptions about the apparent dichotomy between serious and playful context on one hand, and the vague definition of the different types of single combats (including judicial combats) on the other has been stated and calls for further research and case studies.[314]

By cross-analysing different types of sources, such as narrative descriptions, normative or legal sources and technical literature, the historian gains access to relevant information about the training (time spent, type of teaching, actors, even monetary value of the teachings), the rituals and material culture associated with these combats (procedures, gestures, weaponry, clothing) as well as actual details of the fighting itself. In the context of the studies of Historical European Martial Arts—and its main source: the fight books—this type of investigation sheds new light on one of the many contexts of application of the martial gesture. If studied with a praxeological perspective, including comparison with other types of single combat in Western Europe at the end of the Middle Ages, it also sheds new light on the whys and wherefores about ritualised praxes of violence.

Bibliography

Sources (including editions)

Chronique de Georges Chastellain: Les fragments du livre IV révélés par l'Additional Manuscript 54156 de la British Library, ed. by Jean-Claude Delclos, Genève 1991.
Chronique de Mathieu d'Escouchy, ed. by Gaston du Fresne Beaucourt, vol. 2, Paris 1863.
Constitutiones Concilii quarti Lateranensis una cum Commentariis glossatorum (Monumenta iuris canonici. Ser. A, Corpus glossatorum, 2), ed. by Antonius G. García y García, Città del Vaticano 1981.
Das Rechtsbuch Kaiser Ludwigs des Bayern von 1346 (Bayerische Rechtsquellen 4), ed. by Wilhelm Volkert, München 2010.
Fiore dei Liberi, Flos duellatorum : manuale di arte del combattimento del XV. secolo, ed. by Marco Rubboli/Luca Cesari, Rimini 2002.
French, Thomas, York Minster: The St. William Window (Corpus Vitrearum Medii Aevi Great Britain Summary Catalogue 5), Oxford 1999, pp. 70-73.

---

[314] See for example JASER, 2016, pp. 221-242 and JAQUET, 2016b.

Hans Talhoffer, [Königsegg treatise], 1446-1459. Königsegg-wald, Gräfliches Schloss, Hs. XIX 17-3. ed. by André Schulze, 2 vols, Mainz 2010. See App. 1.

Hans Talhoffer, compendium (*Alte Armatur und Ringkunst*), 1459. Kobenhavn, Det Kongelige Bibliotek, Thott 290 2°. ed. by Dieter Bachmann, s.d., online: http://schwert-fechten.ch/quellen/hans-talhofer/talhofer-1459/, accessed 03.10.2015. See App. 1.

Hans Talhoffer, compendium, 1448. Gotha, Forschungsbibliothek, Hs Chart. A558.

Les mémoires de messire Olivier de la Marche (Collection complete des mémoires relatifs à l'histoire de France, 10), ed. by Claude B. Petitot, vol. 2, Paris 1825.

Oeuvres de Georges Chastellain, ed. by Joseph Kervyn de Lettenhove, vol. 3, Brussels 1864.

Paulus Kal, compendium (*Allerley Kempf zur Rosz vnd Fuesz jn vnnd on harnisch*), ca. 1480. Wien, Kunsthistorisches Museum, KK5126. ed. by Carsten Lorbeer et al., 2006, online: http://www.pragmatische-schriftlichkeit.de/transkription/edition_paulus_kal.pdf, accessed 03.10.2015. See App. 1.

Regesten der Grafen von Katzenelnbogen: 1060—1418, ed. by Karl E. Demandt, 4 vols., Wiesbaden 1953.

Schwabenspiegel: Kurzform. 1. Landrecht. 2. Lehnrecht (Monumenta Germaniae Historica. Fontes iuris Germanici antiqui, Nova series), ed. by Karl A. Eckhardt, vol. 4, 1-2, Hannover 1972.

The Book of Chivalry of Geoffroy de Charny: Text, context and translation, ed. by Richard W.Kaeuper/Elspeth Kennedy, Philadelphia 1996.

Traicté de la forme et devis comme on faict les tournois par Olivier de la Marche, Hardouin de la Jaille, Anthonie de la Sale, etc, ed. by Bernard Prost, Paris 1878.

Urkundenbuch zur Geschichte der Herren von Hanau und der ehemaligen Provinz Hanau, ed. by Heinrich Reimer, vol. 3, Leipzig 1892.

Willems, Jan Frans, De Leuvensche kampvechter ten jare 1236, in: Belgisch museum voor de Nederduitsche taelen letterkunde en de geschiedenis des vaderlands, vol. 1, Gent 1837, pp. 26-32.

William Gergory's Chronicle of London, in: The Historical Collections of a Citizen of London in the Fifteenth Century (Camden Society, New Series, 17), ed. by James Gairdner, London 1876.

Zwickauer Rechtsbuch (Germanenrechte NF, Abt. Stadtrechtsbücher), ed. by Hans Planitz/ Günther Ullrich, Weimar 1941.

Secondary Literature

Anglo, Sydney, The Martial Arts of Renaissance Europe, New Haven 2000.

Berguerand, Claude, Le duel d'Othon de Grandson (1397). Mort d'un chevalier-poète vaudois à la fin du Moyen Âge, Lau-sanne 2008.

Boffa, Sergio, Les manuels de combat ("Fechtbücher" et "Ringbücher") (Typologie des sources du Moyen Âge occidental, 84), Turnhout 2014.

Burkart, Eric, Die Aufzeichnung des Nicht-Sagbaren. Annäher-ung an die kommunikative Funktion der Bilder in den Fechtbüchern des Hans Talhofer, in: Das Mittelalter, 19/2 (2014), pp. 253-301.

Carbasse, Jean-Marie, Le duel judiciaire dans les coutumes mé-ridionales, in: Annales du Midi, 87/124 (1975), pp. 385-403.

Cauchiès, Jean-Marie, Duel judiciaire et "franchise de la ville". L'abolition d'une coutume à Valenciennes en 1455, in: Mélanges Fritz Sturm, ed. by J.-F. Gerkens, Liège 1999, pp. 655-668.

Cavina, Marco, Il duello giudiziario per punto d'onore. Genesi, apogeo e crisi nell'elaborazione dottrinale italiana (secc. XIV-XVI), Torino 2003.

Cavina, Marco, Il sangue dell'onore: Storia del duello, Roma 2005.

Contamine, Philippe, La guerre au Moyen Âge, 6th ed., Paris 2003.

Coulin, Alexander, Der gerichtliche Zweikampf im altfranzös-ischen Prozess, Berlin 1906.

Deruelle, Benjamin/Gainot, Bernard (eds.), La construction du militaire. Volume 1: Savoirs et savoir-faire militaires à l'époque moderne, Paris 2013.

Elema, Ariella, Trial by Battle in France and England, unpubl. P.hD. dissertation, University of Toronto 2012.

Fontaine, Marie-Madeleine, Le condottiere Pietro Del Monte, Paris 1991.

Forgeng, Jeffrey L., Pietro Monte's Exercises and the Medieval Science of Arms, in: The Armorer's Art. Essays in honour of Stuart Pyhrr, ed. by Donald J. La Rocca, Woonsocket 2014, pp. 107-114.

Fortner, Sarah, "Kempflich angesprochen" über Kampfgerichte und Kampfrecht, in: Mittelalterliche Kampfesweisen, Bd. 3: Scheibendolch und Stechschild, ed. by André Schulze, Mainz 2007, pp. 10-22.

Galas, Matt, Statutes of the Fencing Masters of Bruges (1456), in: Arts de combat. Théorie & Pratique en Europe—XIVe-XXe siècle, ed. by Fabrice Cognot, Paris 2011, pp. 137-152.

Gevaert, Bert and van Noort, Reinier, Evolution of Martial Tradition in the Low Countries: Fencing Guilds and Trea-tises, in: Late Medieval and Early Modern Fight Books. Transmission and Tradition of Martial Arts in Europe (14th-17th Centuries), ed. by D. Jaquet, K. Verelst and T. Dawson, Leiden 2016, pp. 376-409.

Guenée, Bernard, Comment le Religieux de Saint-Denis a-t-il écrit l'histoire? L'exemple du duel de Jean de Carrouges et Jacques le Gris (1386), in: Pratiques de la culture écrite en France au XVe siècle (Textes et études du Moyen Age 2), ed. by Monique Ornato/Nicole Pons, Louvain-la-Neuve 1995, pp. 331-343.

Hiltmann, Torsten, Spätmittelalterliche Heroldskompendien: Referenzen adeliger Wissenskultur in Zeiten gesellschaftlichen Wandels (Frankreich und Burgund, 15. Jahrhundert) (Pariser historische Studien 92), München 2011.

Hiltmann, Torsten/Israel, Uwe, "Laissez-les aller." Die Herolde und das Ende des Gerichtskampfs in Frankreich, in; Francia, 34/1 (2007), pp. 65-84.

Israel, Uwe, Wahrheitsfindung und Grenzsetzung. Der Kampfbeweis in Zeugenaussagen aus dem frühstaufischen Oberitalien, in: Quellen und Forschungen aus italienischen Archiven und Bibliotheken, 88 (2008), pp. 121-147.

Israel, Uwe/Ortalli, Gherardo (eds.), Il duello fra medioevo ed età moderna: prospettive storico-culturali (I libri di Viella 92), Roma 2009.

Jaquet, Daniel, "Personne ne laisse volontiers son honneur être tranché." Les combats singuliers "judiciaires" d'après les livres de combat, in: Armes et jeux militaires dans l'imaginaire (XIIe-XVe siècles), ed. by Catalina Girbea, Paris 2016b (forthcoming).

id., Le geste, le mot et l'image: La mise par écrit de l'art du combat à la fin du Moyen Âge (De Diversis Artibus, n.), Turnhout 2016a (forthcoming).

id., Fighting in the Fightschools late 15c., early 16c., in: *Acta Periodica Duellatorum*, 1 (2013), pp. 47-66.

Jaser, Christian, "Infamis etiam campio non esse potest." Kämpen in deutschen und italienischen Städten des Spätmittelalters zwischen Marginalität und Rechtspflege, in; Das Mittelalter, 19/2 (2014), pp. 380–406.

Jaser, Christian, Ernst und Schimpf. Fechten als Teil städtischer Gewalt- und Sportkultur, in: Agon und Distinktion. Soziale Räume des Zweikampfs zwischen Mittelalter und Neuzeit, ed. by U. Israel and C. Jaser, Berlin 2016, pp. 221-242.

Jaser, Christian/Israel, Uwe, Einleitung. Ritualisierte Zweikämpfe und ihre Akteure, in: Das Mittelalter, 19/2 (2014), pp. 241–248.

Jezler, Peter, Gesellschaftsturniere—Die Turnierhöfe der deutschen Ritterschaft im Spätmittelalter, in: Ritterturnier. Geschichte einer Festkultur, ed. by Peter Jezler et al., Luzern 2014, pp. 57-72.

Keen, Maurice H., Chivalry, New Haven/London 1984.

Lawrence, David, The complete soldier (History of Warfare 53), Leiden 2009.

Lecuppre-Desjardin, élodie, Le duel judiciaire dans les villes des anciens Pays-Bas bourguignons: privilège urbain ou acte de rébellion?, in: Agon und Distinktion. Soziale Räume des Zweikampfs zwischen Mittelalter und Neuzeit, ed. by U. Israel and C. Jaser, Berlin 2016, pp. 181-198.

Leiser, Wolfgang, Süddeutsche Land- und Kampfgerichte des Spätmittelalters, in; Württembergisch Franken, 70 (1986), pp. 5-17.

Leng, Rainer et al., Katalog der deutschsprachigen illustrierten Handschriften des Mittelalters Band 4/2, Lfg. 1/2: 38: Fecht- und Ringbücher, München 2008.

Ludwig, Ulrike et al. (eds), Das Duell—Ehrenkämpfe vom Mittelalter bis zur Moderne (Konflikte und Kultur—Historische Perspektiven 23), Konstanz 2012.

McAuley, Finbarr, Canon Law and the End of the Ordeal, in: Oxford Journal of Legal Studies, 26/3 (2006), pp. 473-513.

Morel, Henri, La fin du duel judiciaire en France, in: Mélanges Henri Morel (Collection d'histoire des idées politiques 2), 1st publ. 1964, Aix-en-Provence 1989, pp. 175-243.

Neumann, Sarah, *Der Gerichtliche Zweikampf: Gottesurteil, Wettstreit, Ehrensache (Mittelalter-Forschung, 31)*, Ostfildern 2010.

Norton, Christopher, *St. William of York*, York 2006.

Nottarp, Hermann, Gottesurteilstudien, Muenchen 1956.

Oschema, Klaus, Toucher et être touché: l'emploi de gestes dans les batailles judiciaires et le façonnement des émotions dans la résolution des conflits, in: Médiévales (La chair des émotions. Pratiques et représentations corporelles de l'affectivité au Moyen Age), 61/2 (2011), pp. 142-161.

Paravicini, Werner, Ein berühmter Fall neu betrachtet: das Gerichtsduell des Jean de Carrouges gegen Jacques Le Gris von 1386, in: Agon und Distinktion. Soziale Räume des Zweikampfs zwischen Mittelalter und Neuzeit, ed. by U. Israel and C. Jaser, Berlin 2016, pp. 23-84.

Rühl Joachim K., Regulations for the Joust in Fifteenth-Century Europe: Francesco Sforza Visconti (1465) and John Tiptoft (1466), in: The International Journal of the History of Sport, 18/2 (2001), pp. 193-208.

Russell Michael J, The champion's master in trial by battle: a note, in: Journal of Legal History, 5/1 (1984), pp. 76-78.

Russell Michael J, Accoutrements of Battle, in: Law Quarterly Review, 99/3 (1983), pp. 432-442.

Stangier, Thomas, "Ich hab herz als ein leb..." Zweikampf-realität und Tugendideal in den Fechtbüchern Hans Talhof-fers und Paulus Kals, in: Ritterwelten im Spätmittelalter (Schriften aus den Museen der Stadt Landshut 29), ed. by Franz Niehoff, Landshut 2009 pp. 79-93.

Telliez, Romain, Preuves et épreuves à la fin du Moyen Âge. Remarques sur le duel judiciaire à la lumière des actes du Parlement 1254-1350, in: Hommes, cultures et sociétés à la fin du Moyen Âge: Liber discipulorum en l'honneur de

Philippe Contamine (Cultures et civilisations médiévales 57), ed. by Patrick Gilli/Jacques Paviot, Paris 2012, pp. 107-121.

Vones Ludwig, Un mode de résolution des conflits au bas Moyen Age: le duel des princes, in: La Guerre, la violence et les gens au Moyen Age, I: Guerre et violence, ed. by Philippe Contamine/Olivier Guyojeannin, Paris 1996, pp. 321-332.

Welle Rainer, "...und wisse das alle höbischeit kompt von deme ringen" der Ringkampf als adelige Kunst im 15. und 16. Jahrhundert: eine sozialhistorische und Bewegungsbiographische Interpretation aufgrund der handschriftlichen und gedruckten Ringlehren des Spätmittelalters (Forum Sozialgeschichte 4), Pfaffenweiler 1993.

Zabinski, Grzegorz, The Longsword Teachings of Master Liechtenauer: The Early Sixteenth Century Swordsmanship Comments in the «Goliath» Manuscript, Torun 2010.

Appendix 1. Edition of sources

(For references and transcription norms, see Acknowledgment" at the end of the Appendix)

A. About the kampf

Hans Talhoffer, "Von dem kempfen", in Hans Talhoffer compendium (*Alte Armatur und Ringkunst*), 1459. Kobenhavn, Det Kongelige Bibliotek, Thott 290 2°, ff. 8r-10v.

A1. [8r] hie vint man geschriben von dem kempfen
Item wie daz nun sy daz die die decretaleß kempf verbieten, So hat doch die gewonhait herbracht von kaisern und künigen fürsten und hern noch gestatten und kempfen laussen, und darzu glichen schierm gebent, und besunder und umb ettliche sachen und artikeln, alß her nach geschriben staht. Item zu dem ersten maul daz Im nymant gern sin Eer laut abschniden[a] mit wortten ainem der sin genoß ist Er wolte Er hebat mit im kempfen wie wol er doch nit recht wol von Im kem ob er wölte und darumb so ist kámpfen ain mútwill
A2. Item der sachen und ardickelen sind siben Darumb man noch pfligt zu kempfen:
Item daz erst ist mortt
Daz ander verråtterniß
Das dritt ketzerÿ
Daz vierd wöhher an sinem herrn trulos wirt
Daz fünfft umb fanknuß in striten oder sunßt
Daz sechst umb valsch
Daz sibent da ainer junckfrowen oder frowen benotzogt
A3. Item spricht ain man den andern kempflich an, der sol komen für gericht und sol durch sinen fürsprechen sin sach für legen, darumb er in denn an kagt und sol den man

nennen mit dem touff namen und zů namen. So ist recht, daz
er in fůr gericht lad und in dry stund beclag uff dryen
gerichten nach ain ander kumpt er denn nit und veranttwurt
sich nach nymant von sinen wegen, so mag er sich fůrbaß nit
mer veranttwurten, [8v] er bewyse dann Ehafte nott als recht
sy, so sol man in verurtailen alß fer in daz sin bott innerhalb
landes begriffen hant. Je dar nach, alß die ansprach ist
gegangen, darnach sol daz urtail ouch gan.

A4. Item der da kempflich angesprochen wirt uff den dryen
gerichten und er ainost zů der antwort kumpt und legnot
darumb man in an gesprochen hat und spricht er sy des also
unschuldig und der sag uff in daz nit war sy und daz wŏll er
widerumb mit kempfen beherten und uff in daz wysen alß
denn recht sy un dem land darinn eß sy und forttert dar über
mit urtail seinen lertag, so werdent im sechß wochen ertailt zu
sinem lertag und vier tag von dem gericht werdent im auch
ertailt, daruff sie kempfen sůllent alß in dem land gewonhait
und recht ist. Item versprechent sich zwen man willkůrlich
gen einander ain kampfez vor gericht, den git an auch sechs
wochen lertag und sol in frid bannen baiden, und wolcher
under den den frid brech, uber den richtet man on den kampf
alß recht ist.

A5. wie ainer dem anderen mit recht uß[b] gan mag

Item ist daz ein man kempflich angesprochen wiert von aim
der nit alß gůt ist alß er, dem mag er mit recht uß gan ob er
wil oder ob ain man echtloß gesagt wůrde oder worden wer,
dem mag man ouch des kampfes absin. Item spricht aber der
edler den mindern an zu kempfen, so mag der den minderen
nit wol absin.

A6. [9r] Item wie aber zwen mann nit mit ainander můgent
kempfen und wolcher wil under den zwayen dem andern wol
uß gan mag

Item wenn zween mann gesinnt sind biß uff die fůnffte sipp
oder nåher die můgent durch recht nit mit ein ander kempfen
und des můssen siben mann schwern die vatter und můtter
halb måge sind.

A7. Item wie aber ainer dem anderen kampfes absin mag mit
solichem gelimpf alß hie geschriben ståt

Item ob ain lamer man oder einer der bŏse ougen hett und
kampfes an gesprochen wirt der mag sich der auch wol
behellffen und dem gesunden ußgan, eß sy denn daz wyse lůt
daz gelich nach der person machen und daz můssent wyß lůt
uff ir eid tun und daz also glich machen. Es mag auch der lam
oder mit den bosen ougen wol ainen an ir statt gewinnen der
fůr iro ainen kempfe.

A8. Item wenn also die sechß wochen uß sind und der letst
tag komen ist den in der richter beschaiden haut daruff
kempfen sullen, so sullen sie beide fůr den richter komen mit
solichem ertzŏgen und in solich acht alß die gewonheit und
das recht lert in dem lande dar inn sie kempfen sullen oder
nach dem alß sie mit ainander gewillkůrt habent. Item etc.

A9. Item so soll da der cleger schweren daz er der sach darumb er dem ainen man zugesprochen haut schuldig sy und denn so sol man in ainen ring machen und grieß wartten und urttail geben [9v] nach wyser luta raut und nach des landeß gewonhait. Und wer uff den tag in den ring nit kumpt den urttailt man sigeloß in irre denn Ehafte nott die sol er bewysen alß recht ist –

A10. Hie staut wie man sich halten sol wenn die kempfer in dem ring komen sind uff die stund und uff die zit so man pheindiglich kempfen sol WEnn die kempfer also in den ring komen sind So sol der richter von stund an alle stúr und ler vestecklich verbieten by lyb und gůt und sol nicht gestatten daz man einem fúr den andern nicht zulege und sel inß beiden machen so er imer gelichest mag ungenerde.

A11. Das ist was recht wer ob der kempfer ainer uss dem ring fluch oder getriben wurd

Item wolcher kempfer uss dem ring kumpt Ee denn der kampf ain ende haut Er werde daruß geschlagen von dem andern oder fluche daruß oder wie er daruß kâme oder aber ob er der sache vergicht darumb man in denn mit recht an gesprochen haut, den sol man sigeloß urttailen. Oder wolcher den andern erschlecht und ertôtett der haut gesiget. Dem sol man aber richten alß des landes gewonhait und recht ist darumb sie dem mit ainander gekemppffet hand.

A12. [10r] Nun merck uff dissen punten der ist notturfftlich zů uerstend

Item des ersten so soltu den maister wol erkennen der dich lerren wil dz sin kunst recht und gewer sy und dz er frum sy und dich nit veruntruwe und dich nit verkúrtz in der lerr und wiß die gwer zů zerbraitten da mit er kempfen wil. Och sol er den maister nit uff nemen er schwer im dann sin frumen zwerbent und sin schaden zwendent deß glich sol er dem maister wider umb sweren sin kunst nit witter zleren.

A13. Hie merck uff den maister

Item der maister der ain understat zu leren, der sol wißen daß er den man wol erken, den er lerren wil, ob er sie schwach oder starck, und ob er gâch zornig sÿ oder senftmůttig, och ob er gůtten auttem hab oder nit, och ob er arbaitten můg in die in die harr; und wenn du inn wol erkunet haust in der lerr, un wz arbait er uermag dar nach můstu in lerren dz im nůtz ist gen sinnen vind. Och sol der kempffer und der maister sich hůtten dz sie niemand zu sehen laussend und in sunder sie gwer da mit sy arbaittent und sich baid hůtten vor vil geselschafft und von dem vechten wenig sagen dz kain abmercken da von kom.

A14. von kuntschafft

wie der kempffer und der maister kuntschafft môchte hon zu rem widertail, wz sin wesen wer, ob er sy strarck oder swach, ob er och sy gechzornig oder nit, und wie sin touff nam hieß, ob man wôlt dar uß bracticiern oder rechnen. Es ist och nottúrfftig zu wissen wz maister in lerr dz man sich dar<u>nach</u> múg richten.

A15. wenn er nun gelert ist und in den schrancken sol gon

So sol er zu dem ersten bichten, dar nach sol im ain priester ain meß lesen von unßer frowe<u>n</u> und von sant Jörgen, und der priester sol im segne<u>n</u> sant Johanns myn<u>e</u> und de<u>m</u> kempfer geben. Dar nach sol der maister in ernstlich <u>ve</u>rsúchen [10v] und inn under richten dar uff er bliben sol, und sol in uff kein ding haissen acht hon dann uff sin vind, und den ernstlich an schowen.

A16. Merck uff dz infúren

Item wenn der man kompt in den schrancken so sol er machen mit dem rechten fuß ain krútz und mit der hand ains an die brust und sol fúrsich gon im name<u>n</u> des vatters und suns und des hailigen gaists. Dann sind in die grieß<u>wartten</u> nemen und sind inn fúrren gegen der sunne<u>n</u> umbhe. So sol dann der kempfer die fúrsten un<u>d</u> herre<u>n</u> bitte<u>n</u> und die[c] um<u>b</u> den kraiß sta<u>nd</u> dz sy im wólle helffen got bitten dz er Im sig wólle geben gegen sinem vind und alz er war und recht hab.

A17. Dar nach sol man in setzen in den sessel

Wenn er nu<u>n</u> gesessen ist so soll man im fúrspanne<u>n</u> ain túch und sin bar hinder im an den schrancken und sine gwer sind wol gehenckt sin und gericht nach nottúrfft

A18. Die grieß wartten od<u>er</u> tåpffer

Der maister und die grieß wartten sóllend mercken uff den richter oder uff den, der den kampff an lauffen wirt. Wann der rúfft zu dem ersten mal, so sol er den man haisen uff ston un dz túch von im ziehen, und wann man rúfft zú dem dritten mål so sol er in haissen hin gon und in got enpfelhen.

A19. Von dem nach richter

Item der kempffer sel wartten das im nútzit an dem lib úb<u>er</u> den ring oder schra<u>n</u>cken uß gang dann wz dar úber kem: so stat der nach richter an dem schrancken der hott imß ab mit recht ob er angerúft wirt.

a. ab added in superscript.

b. corrected in superscript above a non legible word.

c. inserted in superscript

B. Prologue to the Königsegg treatise

Hans Talhoffer, Prologue, in [Königsegg treatise], 1446-1459. Königseggwald, Gräfliches Schloss, Hs. XIX 17-3, fol. 1r.

B1. [1r] Item Es ist zú wissent des ersten wen ain býder man zeschaffen haut das Im geschriben wúrt zú dem ernst oder er aim schribt so sol er gedencken das er stelle nach aim maister der in zú dem kampff versorge<u>n</u> kend[a] vnd sol Im den maister haissen geloben das er Im trŵlich sein kunst mittaill Vnd sein haimlichait nit sag vnd auch nit wider in sý das er die kunst niemat wissen lauß die er In lere<u>n</u>

B2. Item Es sol auch der Junckher sich hůtten das er nit vill gehaimß mit den lůtten hab das sein haimlichait niemen erfar vnd das im nit werd v̲ergeben vnd besunder so sol er alltag frů vff stån vnd hŏren ain meß vnd dar nach hain gån vnd sol essen ain schnÿtte santj Johans brott vnd sich arbaitten zwů stund in der ler vnd nit vil faists dings essen Vnd nach mittag aber zwů stund Vnd zenacht so er schlauffen wil gån so sol er essen ain Ruggj schnÿtte brot vß ainem kaltten wasser das macht Im gůtten autem vnd wit vmb das hertz

B3. It̲em wer den das der sŏltte fůr sich gån So sol erschriben In ai̲n stat die Im den dar zů gefelt vmb in lauß vnd vmb glichen schirm vnd wen Im das zů geseÿt wůrt so sol er begern das man Im ain frÿeß gelaitt geb fůr sich selb vnd alle dÿ die mit Im dar koment

B4. It̲em Es sol auch der schirm maister den Junckher niemen der da kempffen will vnd sol In fůren ain haimlich stat als in ain kůrchen vnd sol in haissen nyder [knien][b] vnd got bitten das er Im v̲erlich ain gluckhafft stund vnd[c] Im v̲erlich sůg das er seinem feind angesůg

B5. [Vnd ain gut herez vnd starck fewst hab das ist auch fast gut dar zu][d]

a. i corrected with e.
b. barely legible because of a fold in the parchment.
c. barely legible because of a fold in the parchment.
d. variant in the later copies Wien, Kunsthistorisches Museum, KK5342, fol. 1r and Augsburg, Universitätsbibliothek, Cod.I.6.2°.1, fol. 2r.

C. Advices for the master accompanying his client to the *kampf*.

Anonymous, s.t., in Paulus Kal compendium (*Allerley Kempf zur Rosz vnd Fuesz jn vnnd on harnisch*), ca. 1480. Wien, Kunsthistorisches Museum, KK5126, ff. 128v-129r.

C1. [128v] Tout zw dem erstn sol im sein fursprech wandel dingenn vnd alle recht die ein chempfer von rechtz wegenn habenn sol es sey warner lůsner grieswartl vnd was ein chempfer habenn sol.

C2. Jtem wenn er sein warn benennt so sol er fragenn wie er warnenn sol das er recht thu vnnd nicht vnrecht.

C3. Jtem er sol auch fragenn ob er vnd der warner wol zw dem chempfer gen můge dy weyl er an seiner rue siczt vnnd mit in ir notturfft redenn.

C4. Jtem er sol aber fragenn ob er ein sig gewinne oder verlornn ob er vnnd sein warner wol zw im gen můgen vnd ir noturfft wol mit ym redenn also so er wider an sein rue chumbt ym chrais.

C5. Jtem wenn er denn grieswertel oder lusner genannt so
sol er in fragenn wie er lůsen Sol das er recht thue vnnd
nicht vnrecht.

C6. Jtem er sol auch fragenn wie er mit der stångenn thun
sol das er recht thue.

C7. Jtem er sol auch fragenn ob ir ainer der stangenn begert
wie er die vntterstossenn sol das er recht thue vnnd nicht
vnrecht.

C8. Jtem er sol auch fragenn was die grieswertel oder lusner
hornn oder sehenn wie sy das furbringen das sy recht thunn
vnnd nicht vnrecht.

C9. Jtem er sol auch fragenn was die grieswårtl ein
můtiglich sagenn als die das gesehenn oder gehort habenn ob
es icht billichenn do bey beleyb.

C10. Jtem er sol auch fragenn ob die griswertel mit ein ander
nicht stóssig sein vnd nit vber ains mochten wern zw sagenn
was dann dy vrtailer gesehenn oder gehort hiettenn ob es nit
billichenn do peleybe das weybe leyb vnd das die auch dar
vmb sagenn sůllenn.

C11. Er sol auch fragenn es sey grieswertl oder vrtailer wie
sy darvmben sagenn sullenn das sy recht thunn das man das
ausfunndig[a] mach.

C12. Es sol auch denn chempfer fragen wellicher ein
anchlager sey ob er icht billichen die wal vnnd die vorfart hab
zu dem chrais zw siczenn mit seinem stull wo er wil.

C13. Er so auch denn chempfer fragenn wie manigen
cholben er habenn sul vnd was er habenn sol zw dem champf.

C14. Er sol auch fragenn ob er ein cholbenn verwůrff der aus
dem chrais chemb was recht werr.

C15. Er sol auch fragenn ob er ein cholbenn verwůrff der im
chrais be lib ob im der nicht, billichenn zw staten wider chem
ob er sein begert.

C16. Jtem er sol fragenn ob sein grieswårtl ycht billichenn
ein cholbenn bey im habenn sull oder eins begert das er im
denn mocht zw pringenn das das [sic] er Recht dar ann tat
vnnd nicht vnnrecht.

C17. Er sol fragn was aus dem chrais chem es wer an
hennden am leyb an ann [sic] fuessenn ain schilt ann
cholbenn was darvmb Recht sey.

C18. Er sol fragenn wie manige stanngenn er begerenn sul
vnd wie offt er den Sig domit verlorn hab[b].

C19. [129r] Er sol fragenn ob man sy nicht billichen
beschauenn sul ob sy nichtz vngleichs oder vnpillichs bey in
hettenn.

C20. Er sol auch fragen ob man die schilt vnnd anndern
Zeug nicht billichenn beschauen sull[c].

C21. Er sol fragn wie offt er sein zue habm sull vnnd wie
lanng.

C22. Er sol fragenn wie er auf stenn sull zw dem champf.

C23. Er sol fragenn ob icht pillichenn beleyb panalenn puntten vnd artikel_en_ als die vor mit recht ertaylt sein wordenn vnd in dem puech geschribenn stenn.

C24. Er sol frag_en_ welcher des chambs der nyderlig wie man zw dem selbm richt_en_ sol er sol fragenn wer richtenn sull.

C25. Er sol fragn ob yemand steurt oder lern_n_it mit wortten oder mit werckenn was darvmb recht sey vnnd ob man das icht billichenn verpiett vnd wie man das verpietenn sull.

C26. Ob auf lauff hinder dem ring geschehe ob das denn chempfernn kain schadn pring_en_ sold[d].

a. "i" with two points above.

b. Four last words reported below the line on the right hand side.

c. Last word reported below the line on the right hand side.

d. Last word reported below the line on the right hand side.

Acknowledgement and transcription norms

The transcriptions are based on the following works: A) Bachmann, s.d., online; B) Schulze, 2010, p. 22; C) Lorbeer/Lorbeer/Meier, 2006, pp. 91-92. All transcriptions in this appendix have been revised by the author and the following norms have been applied: Resolution of the abbreviation by underscoring; <u> and <v> / <i> and <j> / <ß> were reproduced as in the original; the diphtongs /uo/, /ue/, /ae/, /oe/, were marked superscript vowels with the special characters <ů>, <ủ>, <ȯ>, <å>; the uppercase and lowercase were respected.

4. The *Bellifortis* of Konrad Kyeser of Eichstätt

Michael Chidester

In 1459, the manuscript called Thott 290 2° (now stored in Copenhagen) was completed at the request Hans Talhoffer. Most of its teachings related to the myriad forms of fighting that were the trade of a fencing master. However, modern readers paging through it for the first time are quick to notice an anomaly: 36 pages toward the beginning of the manuscript contain fantastic depictions of complex war machines, exotic inventions, and strange scenarios. They tend to conclude that Talhoffer must have been an inventor or technologist, his generation's Leonardo da Vinci. What is not apparent is that these are not creations of Talhoffer at all; they are a remnant of Konrad Kyeser's treatise *Bellifortis*[315] and predate Talhoffer's manuscript by over fifty years.

There has been no substantial work on *Bellifortis* in English, and when I started planning this paper, I believed the same to be true in German. The most recent work that I was aware of was QUARG's edition and WHITE's review of it, and my intent was to write a summary of those works along with a comparison of the illustrations in Copenhagen to those of Göttingen 63.[316] When I discovered CERMANN's monograph, I thought I would summarize that instead. However, it soon became apparent that these are just three links in a chain (or an entire mail shirt), and even confining my attention to the most important of major works, the project mushroomed. In addition to those considerations, my own interests tended to not align with those of most authors, so I had to dig into more obscure literature to tease out the connections I was seeing.

As a result of this process, my intent with this article is now threefold:

1. To give a sense of Kyeser's life and place in history.
2. To briefly outline the origin and evolution of *Bellifortis* and its tradition.

Figure 1: A trebuchet.
Legend: Vienna 3026, fol. 2ʳ.

* I'd like to thank KENDRA BROWN for her help with the Latin text, for reviewing the first draft of this paper, and for her patience while this book consumed my life.

[315] QUARG interpreted the title *Bellifortis* as "The War Hero". Others have offered translations along the lines of "battle force" or "fighting strength". CERMAN turns this into "strong/efficient for the war", but also points out that the French Marshall Boucicault, whose attitude during the Nicopolis Crusade may have earned Kyeser's ire, held the Latin title *comes Bellifortis* (Count of Beaufort); Beaufort means "beautiful fortress". See CERMAN (2013), 12-14.

[316] For this paper, I will be using a scheme of manuscript abbreviations based on CERMANN's; for a complete list, see the Bibliography.

3. To review the contents of the 7-chapter *Bellifortis*, with attention to the material found in Copenhagen.

Literature Review

Most research and literature on *Bellifortis* has been written in German, and only a scant few English resources exist. I will not attempt a comprehensive review of the German literature here, but only hit a few high notes. For a complete literature review, see CERMANN 2013.

The dominant work on *Bellifortis* has long been that of GÖTZ QUARG.[317] In 1967, he published a facsimile, transcription, and translation of Göttingen 63. QUARG'S writings framed much of the discourse over the subsequent half-century. This is an unfortunate starting point for several reasons. The manuscript itself is beautiful, but as we will see, a dead end as far as the *Bellifortis* tradition goes and not even the final form that Kyeser himself settled on.[318] Furthermore, his biographical sketch fits better into the genre of historical fiction, and QUARG made many assertions about Kyeser, such as his vocation as a doctor or his imprisonment in the mountains of Bohemia, that are based on the flimsiest of evidence (or none at all). Finally, the transcription and translation that he provided are riddled with errors which scholars have spent the subsequent decades trying to untangle.

UDO FREIDERICH attempted to expand the conversation in 1995[319] with an edition of two more manuscripts from the same library: Göttingen 64 and Göttingen 64a. He correctly recognized that 64a was the precursor to Göttingen 63 and analyzed their relationship, but was hobbled by the fact that Göttingen 64 is a member of the disordered 7-chapter group so his analysis of that branch and its relationship to the two oldest manuscripts is of limited value. He called many of QUARG's findings into question, but didn't venture his own answers to most of them.

In 2002, RAINER LENG published a colossal survey of the war book tradition from the late Classical to early Modern periods,[320] including a lengthy discussion of *Bellifortis* and its place in the history he was crafting. While he was not shy about criticizing earlier authors, *Bellifortis* was unfortunately not his primary interest and his efforts there were directed at showing that the *Bellifortis* tradition is largely disconnected from the gunmaster books that were his main theme.[321]

---

[317] QUARG.

[318] For an explanation of the different versions of *Bellifortis*, see "Composition and Evolution" (below).

[319] FRIEDERICH.

[320] LENG.

[321] Incidentally, it was this study of war books that lead to LENG's involvement with the catalog of illustrated German manuscripts that he is known for in historical martial arts circles. He was tasked with both war manuals and fencing manuals, which the editorial staff apparently believed were similar enough.

REGINA CERMANN finally sought to correct the record in a monograph published in 2013.[322] She dismantles his biographical and historical narrative and shines the light of history (including more recent discoveries) on the subject, rewriting Kyeser's biography and placing him in a more correct historical context. She also assembles the first complete stemma of manuscript copies, and weaves her way through it as she explains the connections between the different branches and texts, though she is largely unconcerned with discussing the individual machines, devices, and incantations (except for a single spell toward the end which QUARG and other authors had also fixated on). A shorter paper published in 2014 addresses changes in illustrations introduced in the course of the tradition.[323]

The extent to which we have been starved of research in English may be encapsulated by the fact that the most commonly-cited English work on *Bellifortis* in scholarly literature, even today, is a 5-page book review of QUARG published by LYNN WHITE, JR. in 1969.[324] WHITE's review includes a summary of QUARG's findings and a few criticisms of them, but is not (and was not intended to be) a standalone work. Less commonly cited is BERTRAND GILLE's monograph of 1966,[325] which has little to say about Kyeser but a lot to say about his war machines and their place in the history of military engineering (in some ways he excels LENG on this subject).

Apart from these, BRIAN PRICE's 2016 article[326] is aimed at a general audience and breaks no new ground. Other discussion has appeared in literature on the history of magic,[327] a topic that features almost as prominently in *Bellifortis* as does engineering, but these treatments are almost entirely dependent on the scholarship already mentioned. A comprehensive English-language edition of *Bellifortis* is desperately needed, but I have neither the space nor the expertise to provide that here; hopefully, another researcher will pick up the torch.

Figure 2: Portrait of Konrad Kyeser.
Legend: Göttingen 63, 139ʳ.

Who was Konrad Kyeser?

Very little is known about Konrad Kyeser himself. Fortunately, he of-

---

[322] CERMANN (2013).
[323] CERMANN (2014).
[324] WHITE.
[325] GILLE. Originally published in French in 1964.
[326] PRICE.
[327] Including EAMON and LÁNG.

Figure 3: Kyeser's heraldry. Legend: Göttingen 64a, fol. 159ʳ.

fered a few biographical details in his own work, including his own horoscope.[328] From this we can determine that he was born on 26 August 1366 in Eichstätt, an Imperial City and seat of the Prince-Bishopric of Eichstätt.

His parents were Elisabeth and Rüdiger;[329] a *Kieser* is a food inspector, though it's also an archaic term for a more general judge or arbiter. This suggests that he came from a Patrician or lower ministerial family.[330] He provides us with his heraldry in two of his manuscripts,[331] but to date it hasn't been identified with any known Eichstätt family.

Eichstätt was a center of learning in the fourteenth century (though not a university city) and the Dominican monastery offered free public lectures on science and philosophy.[332] CERMANN speculates that during Kyeser's youth in Eichstätt he came to the attention of the Counts of Oettingen, the youngest of whom (Friedrich IV) was installed as Prince-Bishop of Eichstätt in 1383. Most of Kyeser's eventual career would happen in proximity to one Oettingen or another, suggesting some kind of relationship.[333]

In *Bellifortis*, Kyeser often refers to himself as an "exile", perhaps because he left his home at a young age and didn't return for a very long time. In 1390, at the age of 24, he enrolled as a law student the University of Prague. He isn't given a title in the register, but since he was seeking a doctorate, he would have been required to show that he had already received a master's degree elsewhere (which typically took six years).[334] Kyeser does not appear on the law register in any subsequent year, so he seems to have not pursued this study very long.

Admission to universities in the fourteenth century required joining the lower clergy by taking one of the minor orders (acolyte, doorman, exorcist, or reader), and classes were typically conducted in Latin. Being ordained as an exorcist, which included receiving a book of exorcism rituals as a token of this calling, was often a gateway to experiments with the occult and necromancy,[335] subjects that Kyeser would later claim

---

[328] Göttingen 64a, 1r-2r.
[329] Göttingen 63 139r.
[330] CERMAN (2013), 44-45.
[331] Göttingen 63, 159r; Göttingen 64a, 138v.
[332] QUARG, XX.
[333] CERMANN (2013), 51-52.
[334] Ibid., 50. CERMANN was unable to find him at any nearby universities (Bologna, Erfurt, Heidelberg, Padua, and Vienna), or in Prague before or after 1390, but notes that available records are incomplete.
[335] KIECKHEFER (2000), 153-4.

expertise in and include in *Bellifortis*.[336] Treatises of learned magic were available in various libraries at the University of Prague, particularly the personal libraries of many masters, and this may be where Kyeser began accumulating his own copies of magical writings.[337]

Kyeser also seems to have spent time in Padua in his youth, and mentions serving Francesco II of Carrera, though it's unclear if this was in the 1380s when he was seeking his master's degree, or after he left the University of Prague in 1391. CERMANN believes he didn't reach Padua until around 1395, but also notes that the Visconti Library's collection of rare scientific and engineering texts might be another inspiration to begin thinking about composing his own book.[338]

This is perhaps the best point in the narrative to discuss Kyeser's occupation. QUARG assigned Kyeser the vocation of physician (or perhaps army medic) but offers little evidence. Kyeser's statement comparison of cutting and thrusting wounds, which QUARG sees as a sign of experience with combat trauma, is merely a repetition of Vegetius' explanation of Roman sword techniques. QUARG also sees significance in the various body parts being named in some of the magic rituals, apparently failing to realize that the incantations Kyeser records were not his own creations.[339] But however flimsy these arguments may be,[340] the possibility should not be discounted.

Several medical doctors in the fourteenth and fifteenth centuries are known to have fulfilled a double role as court physician and magician or astrologer,[341] which lines up nicely with Kyeser's career trajectory. After Kyeser left the University Prague, he might well have sought medical training at a different university during the missing years in the early 1390s. His own account in 1405 includes a list of nobles that he had served in his career, including Francesco II (mentioned above), Wenzel IV of Bohemia (discussed below), Jobst of Moravia, Stephan III of Bavaria, Wilhelm of Austria, Sigismund of Hungary, and others.[342] Apart from the first two, it is unclear what services he might have provided these princes, so physician is as good a guess as any other.

---

[336] Kyeser's interest in (and access to) magical teachings is apparent not only in the texts he included in his treatise, as detailed in the section "Sources of *Bellifortis*" (below), but also in the way he describes the world. In the epilogue to Göttingen 63 (135rv), he separates the sciences into three groups rather than the usual two: the liberal arts, the exceptive arts, and the mechanical arts. The second category, *ars exceptiva*, was a term introduced in a popular book of ritual magic called *Ars Notoria* ("The Notary Art"), and Kyeser confirms that this was his intended reference by further mentioning not only astrology and alchemy in his lists of sciences, but also theurgy (the invocation of supernatural forces) and the "art of Pythagoras". See LENG, 122-123. For more information about the *Ars Notoria*, see LÁNG, 33-34.

[337] LÁNG, 50.

[338] CERMANN (2013), 53-55.

[339] QUARG, XX-XXI. See "Sources of *Bellifortis*", below, for information about these magical recipes.

[340] WHITE further attempts to bolster them by pointing out that the color green was associated with physicians, and Kyeser's patron Jupiter was also the patron of doctors. See WHITE, 438.

[341] LÁNG, 210-214.

[342] Göttingen 63, 137r.

Most of these men were also involved in what would be a defining event in Kyeser's life: the battle of Nicopolis.[343] Indeed, he could have marched to this war in the service of any of them.

There is no indication of what Kyeser's role in the army was, and QUARG's assertion that he was a mid-level commander[344] is baseless speculation. If he were not a medic, then it's equally possible that he travelled with the army in a clerical or logistical role. That might go some way toward explaining his statement that he had experience in forming a wagon-fort.[345] A non-combat role would also explain how he managed to escape a battlefield in which most of the soldiers were captured or killed.

In any case, he probably returned home some time in 1397, like most uncaptured survivors.

It's unclear whether Kyeser ever saw combat outside of that disastrous campaign. QUARG involves him in the Bavarian War of the Cities (1387-89) in service to Stephan, and possibly even the Paduan campaign of 1390 which restored the city to Francesco.[346] This speculative military career would make it difficult to squeeze in a university education before going to war in 1387 (and require him to rush from Padua to Prague later in 1390), but it's not impossible.

Nevertheless, Nicopolis left its mark on Kyeser, and he came to see it as an object lesson for all the problems in the common mode of warfare. His ideas about the proper conduct of battles, sieges, and other related activities—based on the failures of Nicopolis—ultimately grew into the treatise that is our subject here: *Bellifortis*.

His travels after Nicopolis are unknown, but the *Bellifortis* was certainly the work of a long period of time. Writing Göttingen 64a was probably the work of at least two or three years, as that is the approximate gap between each subsequent edition.[347]

By 1402, Kyeser was living in Kutná Hora (Kuttenberg), a city about 45 miles east of Prague that was a favorite residence of Bohemian royalty. Kyeser seems to have become a member of the court of Wenzel IV (recently-deposed King of the Romans, but still King of Bohemia), perhaps riding the influence of Friedrich III of Oettingen who became the king's chancellor in 1398. Kyeser went so far as to describe himself as the king's confidant, and a "student of occult and natural arts" (*Artis occulta necnon natura scolaris*).[348]

---

[343] This will be discussed in more detail below.

[344] QUARG, XXIII.

[345] Göttingen 64a, 4v.

[346] QUARG, XXI-XXII.

[347] See "Composition and Evolution" (below).

[348] Göttingen 64a, 159r. "Occult" is a potentially confusing word due to modern connotations. In this period, it referred specifically to so-called "natural magic". Natural magic was based on finding the hidden or secret (the literal meaning of "occult") properties of natural objects, as opposed to the talismans and other manufactured magical objects of image magic or the supernatural invocations of necromancy. Magnetism was a popular

In the introduction to Göttingen 64a (presumably the final part written), Kyeser records his observations of a comet visible to the naked eye and specifies that he observed in on 7 March 1402. Wenzel had been captured by his half-brother Sigismund of Hungary just one day earlier,[349] and in this light Kyeser sees the comet as an ill omen and a sign that righteous men should rise up to support his master as the rightful Holy Roman Emperor and protector of their people.[350]

On 23 June 1405, he finished Göttingen 63, a revised and enhanced version of the treatise. By this time, Kyeser was living near Žebrák Castle, about 30 miles southeast of Prague. He also seems to have married, though the identity of his wife is unknown,[351] and may have even had a son, named Rüdiger after Kyeser's father.[352]

The new manuscript contained his famous polemic against Sigismund, replacing talk of comets and supernatural omens with a reminder of the Battle of Nicopolis and Sigismund's alleged cowardice (even though Kyeser acknowledges that he also fled the field[353]). Despite these accusations, the true source of Kyeser's hate for Sigismund seems to be his kidnapping and imprisonment of Kyeser's overlord.

Göttingen 63 is much more artistically ambitious than the first one, employing German artists from the Prague school and Italian scribes. It was dedicated to Ruprecht, the new King of the Romans (though internal evidence suggests that it was initially planned to be dedicated to Wenzel again).[354] At the end of the manuscript, he included his own horoscope which indicated among other things that he expected to go on a journey in his

Figure 4: Heraldry of Kyeser and his wife.
Legend: Göttingen 63, fol. 138ᵛ.

---

example of natural magic to justify the practice. Medieval church authorities were divided over whether natural magic was inspired by the devil or should be allowed, but some, including William of Auvergne and Pseudo-Albertus Magnus, defended the practice. The phrase "natural and occult arts" expresses an interest in both the physical sciences and natural magic. For a discussion of categories of magical practice and their status in society, see LÁNG, 24-43.

[349] QUARG imagined that Kyeser was captured along with Wenzel and devised all his subsequent writings as he slowly died in prison, not surviving long after the completion of his second manuscript in 1405. At one point he suggests an even more fantastic picture, of Kyeser writing in secret and passing scraps of paper for his lieutenants to smuggle out of the prison and have written into a book. Painters who happened to be passing through Schaunberg were conscripted to illustrate his writings without his supervision, and so on. See QUARG, XXIII-XXV. There is no evidence to support this narrative, but the evocative fantasy appears frequently in subsequent literature.

[350] Göttingen 64a, 1r-2r.

[351] Göttingen 64a presents his heraldry in isolation on 159r, but on 138v of Göttingen 63, it is joined by a second heraldic achievement facing his. This presentation is a typical when two people of high status marry.

[352] LENG, 110.

[353] Göttingen 63, 3r.

[354] CERMANN (2013), 43.

Figure 5: Welcoming Kyeser upon his return home to Eichstatt?
Legend: Rome 1994, fol. 69ᵛ.

40th year, 1406.[355] This might be a reference to Friedrich III of Oettingen's move to Ruprecht's court and a hope that he would also find a place there.[356]

Very little is known of Kyeser after this time. He left space in Göttingen 63 for his death date to be recorded,[357] but it was never filled in.[358] If he made the journey to Germany in order to present himself to Ruprecht in 1406, he did not find the reception he hoped for. There is no record of him there, and his manuscript never entered the Palatine library. He may have subsequently attempted to sell his book to the city of Nuremberg, though that sale also never took place.[359]

CERMANN suggests that this series of rejections was the political consequence of his attacks on Sigismund,[360] but other authors offer a more exotic (if less credible) explanation: Kyeser's reputation as a sorcerer had grown too great and people now feared him.[361]

CERMANN proposes that in the end he settled in Eichstätt, the "exile" returning home.

The only other evidence of his activities after 1405 is further revisions to *Bellifortis*. A third version seems to have been prepared in 1407, this time translated to German (only fragmentary copies of this version exist).

In 1410, at the age of 44, he reorganized the contents of the treatise and reduced the number of chapters from ten to seven. These efforts might have been attempts to broaden the appeal of the work after it had been ignored by the high and mighty.

Finally, another 7-chapter manuscript was created in the 1410s with experimental additions (such as illustrations for previously-unillustrated text); this is not a sure sign of Kyeser's involvement, but he would have only been in his 40s at this point and his involvement shouldn't be ruled out.[362]

The Battle of Nicopolis

Many books have been written about this important battle (and campaign), and it's impossible to do justice to it here. I

---

[355] Göttingen 63, 139rv.
[356] CERMANN (2013), 68.
[357] Göttingen 63, 137r.
[358] QUARG kills Kyeser at this point in the narrative, possibly before Göttingen 63 was even completed, but as with his other biographical elaborations, he offers no evidence in favor of this end.
[359] CERMANN (2013), 68.
[360] Ibid.
[361] EAMON, 191.
[362] CERMANN (2013), 82.

will offer a brief description, though, since so many details of this battle find echoes in *Bellifortis*.

In the late fourteenth century, an expansionist Ottoman Empire had besieged Constantinople and pushed the borders of Christian Europe to the edge of Hungary, threatening wide swathes of Eastern Europe. In 1394, at the request of Sigismund of Hungary, Pope Boniface IX called for a crusade. This call didn't immediately produce results; many considered the days of crusades against the east to be long past, and the papacy was presently in the middle of a schism, with Antipope Clement VII opposing Boniface from Avignon.

Figure 6: Sigismund welcomes the French crusaders to Buda. Legend: Sebastien Mamerot, 1473; Paris, Bibliothèque nationale de France, Français 5594, fol. 260ʳ (detail).

However, a lull in the Hundred Years War had stretched on for several years and the Phillip II "the Bold", Duke of Burgundy, was looking for new avenues for his knights to gain glory. In 1395, he notified Sigismund that a request sent to the King of France (his nominal overlord) would be accepted.

Numbers are impossible to nail down since accounts vary so wildly, but at least 5,000 French knights rode to the war in 1396, as well as at least 6,000 mounted archers and infantry, all under the command of Phillip's eldest son John, Count of Nevers. The Venetian navy carried them to Buda, and there they joined a coalition of Eastern European forces and Knights Hospitaller lead by Sigismund, King of Hungary. The total strength of the Crusader army has received wildly different estimates from various authorities over the centuries, ranging from 16,000 to as much as 130,000.

Assembling in Hungary took much of the summer, but ultimately the leaders all arrived and held a council of war. Sigis-

Figure 7: Sigismund's war council. Legend: Jean Froissart, 1470-2; London, British Library, Harley MS 4380, fol. 84ʳ (detail).

mund proposed that they wait and allow the Ottomans to march to Buda (as they had announced they would). The French leaders rejected this plan, calling it cowardly, and insisted they ride out to find them in the field. Sigismund acquiesced, and finally the combined armies marched south along the Danube to the Iron Gates gorge, where they crossed the river on boats (a process that took eight days). This took them into occupied territory, and they began pillaging the countryside and sacking cities, including conquering and taking prisoners at Oryahovo. On 12 Septem-

ber, they arrived at the fortress of Nicopolis.

Nicopolis had laid in supplies and was well-defended, with high walls and terrain advantages including cliffs that limited approach to a single steep slope. The crusader army was not prepared to for siege warfare and had brought no siege equipment; their only option was to attempt to starve the city. Boucicault, one of the French commanders, opined that ladders were easy to build and other siege engines were worthless compared to bold knights.

The Ottomans responded to this incursion by lifting the siege of Constantinople and riding to Nicopolis, arriving on 25 September. The French knights, frustrated by two weeks of inaction, demanded to lead the first assault over the objections of Sigismund and his commanders, all of whom had experience with Ottoman tactics. When the army was a few hours away, the French also decided to slaughter the prisoners they had taken at Oryahovo (which horrified the Hungarians and was later described by their own chroniclers as barbaric).

Accounts of the battle itself are varied. It seems clear that rather than allowing the Ottomans to advance to meet them, the French knights took their mounted archers and rode to engage them where they camped in the distant hills. The first group of soldiers they encountered were ineffective peasant conscripts (as Sigismund had predicted to them). More problematic were the covering fire from hidden archers and the rows of sharpened stakes arranged along the hills. Still, this advance force was routed rather quickly, though many French horses were lost to the stakes and arrows and many French knights were forced to dismount and pull the stakes up by hand.

Despite being warned that the first line of infantry would be fodder, the French forces seem to have believed it was the main body of the army that had broken. Even some of the French commanders wanted to pause and regroup at this point so their infantry and Sigismund's army could catch up, but the younger knights refused and pressed on, thinking to pursue and slaughter a retreating enemy.

Once the French knights had surmounted the hills and reached the plateau, instead of finding the fleeing remnants of a rout, they encountered the main body of fresh Ottoman soldiers. The French attempted to rally, but winded from the first round of battle and with many of them dismounted, they were completely overmatched. In the end, the survivors

Figure 8: The French knights encounter the Ottoman army. Legend: Sebastien Mamerot, 1473; Paris, Bibliothèque nationale de France, Français 5594, fol. 260ʳ (detail).

surrendered themselves for ransom. His leadership in this dubious strategy and humiliating defeat are what earned the Count of Nevers the nickname John "the Fearless", which he would wear proudly for the rest of his life.

Other units of Ottoman cavalry had swept around the barricade to strike both flanks of the remaining Crusader army under the command of Sigismund, which was hurrying to catch up to the French. Estimates of the total strength of the Ottoman forces that day range from 15,000 to 200,000.[363]

The Romanian contingent reportedly saw riderless horses racing toward them from the direction of the front line (probably escaped from the Ottoman camp), concluded the day was lost, and fled before the Ottomans even attacked. Sigismund hastily laid battle lines to try to prevent the envelopment of his army, but the arrival of 1,500 Serbian heavy cavalry later in the day to support the Ottomans of ended any chance of regaining control of the battle. Sigismund fled the field (possibly forced to do so by his lieutenants) while most of his army was surrounded by the Ottomans and captured.

The bloodshed was not over, though. Bayezid, the Ottoman Sultan, learned of the slaughter of his people from Oryahovo and was enraged. He ordered all the French knights assembled before him and separated out knights who were members of the high nobility as well as knights under the age of 20. These were forced to watch while the rest of the captive knights were stripped, tied up, and executed in groups of three or four at a time from morning until early afternoon, when he was finally satisfied with his revenge. Between 300 and 3,000 knights were thus killed, in addition to the thousands that lay dead upon the field.

Meanwhile, the fleeing soldiers of Sigismund were little better off. They ended up trapped on the shores of the Danube with few ships left to ferry them across and Ottomans at their heels. Some of the ships were overloaded by desperate men and sank; many who couldn't reach a boat tried to swim across the river and drowned. Those soldiers who made it across found themselves in country that had already been pillaged by the fleeing Romanians (half a day ahead of them), and many more starved to death along the route.

Figure 9: Ottomans attack Sigismund's army. Legend: Jean Froissart, 15th c.; Paris, Bibliothèque nationale de France, Français 2646, fol. 255ᵛ (detail).

---

[363] Many pre-20th century commentators and historians assign the Ottomans roughly double the number they assign to the Crusaders, perhaps to excuse the crushing defeat they suffered, but the more recent consensus is that the lowest figures are correct and the two forces were fairly evenly matched.

In retrospect, this utter defeat (along with similar embarrassing battles in the early fifteenth century) was a sign of the times. The dominance of heavily-armored cavalry was not ending quite yet, but the era in which wars could be fought and won by chivalrous adventurers was (if it ever existed at all). Kyeser certainly saw the battle in that light, and the events at Nicopolis provided the template for much of the contents of his *Bellifortis.*

Sources of *Bellifortis*

Though *Bellifortis* is often framed as the original invention of a master engineer (or the flights of fancy of a charlatan), it is the direct product of several recognizable literary and textual traditions, some Medieval and some which stretch back to Antiquity. Though the shadow of Nicopolis can be seen across the whole text, little of its contents is original to Kyeser and his role in its creation seems to be that of compiler and commentator rather than author. Kyeser himself was not shy in acknowledging this, describing himself as "an empty vessel for the knowledge of the philosophers".[364]

He openly named two authorities that he drew on.[365] One is Flavius Vegetius Renatus, a Roman writer from the fourth or fifth century. Vegetius was a government official rather than a soldier, and his treatise *De rei militari* ("On the Military Thing") was mostly a redaction of earlier Roman military theorists, but their works did not survive and his did so he enjoyed enormous fame and influence throughout the Middle Ages. Vegetius' treatise would supply the basic plan for the first five or six chapters of Kyeser's work.

The other authority that Kyeser acknowledged was an Antonius Romanus. He may have been referring to the unknown author of *De rebus bellicis* ("On the Things of War"), another famous Roman treatise from the late fourth or early fifth century.[366] This treatise currently only exists in late copies and it's hard to know how the text evolved during transmission, but the extant fifteenth century copies include fantastical depictions of war machines that may have fired Kyeser's imagination.

Alternatively, this might be a reference to Aegidius Romanus, a thirteenth century theologian and intellectual.[367] He is not remembered today as a military writer, but around 1280 he authored a "prince's mirror" for Phillip IV of France. This was divided into four books, and part of the fourth is dedicated to warfare (based equally on Aristotle and Vegetius).

---

[364] Göttingen 64a, 1v. Even with this statement, Kyeser seems to be echoing Vegetius.
[365] Ibid., 1v.
[366] CERMANN (2013), 28-29.
[367] FRIEDRICH, 10, and LENG, 117. LENG further indicates that this is the first time the arts of war were included in a prince's mirror.

Kyeser also states that he drew upon "other reliable authors" but leaves it to readers to guess which ones (perhaps because his other sources were anonymous to begin with). It is clear, though, that the final chapters of *Bellifortis* are indebted to magical and alchemic literature.

He copies directly from the *Experimenta Alberti* ("Experiments of Albert") of pseudo-Albertus Magnus,[368] as well as its frequent companion text *De mirabilibus mundi* ("On the Marvels of the World").[369] These are late-thirteenth century examples of the *experimenta* genre, a group of natural magic[370] texts that often appear in manuscripts in the company of medical, herbal, and alchemical texts (the distinctions between these categories were often nonexistent for Medieval readers).[371]

Kyeser's primary interest in magical literature seems to have been the manipulation of fire, both for illumination and as a weapon, so he also included almost the entirety of the *Liber ignium* ("Book of Fire") of Marcus Graecus. This text first appears in the fourteenth century (though some have speculated much earlier origins for it) and includes recipes for crafting gunpowder, Greek fire, unextinguishable lamps, and various pyrotechnic effects. Such was Kyeser's interest in a comprehensive discussion of fire that he also seems to have accidentally included a few pseudo-Albertian recipes which were themselves copied from *Liber ignium*.[372]

In addition to these, the text mentions an anonymous (and currently unknown) *Liber amicabilis operis* ("Friendly Book of Works"), the *Liber de ingeniis spiritualibus* ("Book of Wind Devices") of Philo of Byzantium, the 9th century Arab scholar Hunayn ibn Ishaq, and the 2nd century Roman authority Galen, but doesn't elaborate on the exact contributions of these sources.

Discussion of Kyeser's inspirations and direct textual borrowings would not be complete without addressing the artwork of *Bellifortis*, particularly what GILLE has named the *primitives* ("primevals") group.[373] These are a series of anonymous manu-

Figure 10: Recipes from Liber ignium.
Legend: Göttingen 63, 102^r.

---

[368] Albertus Magnus (ca. 1200-1280), the famous theologian and scholar, was familiar with alchemy but never ventured into true magical practice. His fame was sufficient, however, for a number of pseudepigraphers to affix his name to their own writings on the subject.

[369] LÁNG, 72-74

[370] See note 347.

[371] LÁNG, 55-61

[372] LÁNG, 72-74

[373] GILLE, 56-58. GILLE's book was originally written in French, and the use of the cognate "primitives" in the English translation might be puzzling since there's nothing crude or undeveloped about the artwork. I will be using "primeval" in this paper to better capture GILLE's intent.

Figure 11: A demon-faced weapon.
Legend: Vienna 3069, fol. 69ᵛ.

Figure 12: A demon-faced weapon.
Legend: Göttingen 64a, fol. 5ᵛ.

scripts that appeared in the early- to mid-fifteenth century (beginning with Vienna 3069 in 1411) which present uncaptioned illustrations that depict many of devices from *Bellifortis* in great detail—in some cases, more detail than Kyeser himself offers.[374] Though Vienna 3069 appears a few years after Kyeser's work, the primevals group is believed to derive from a now-lost work from the fourteenth century.[375]

Vienna 3069 does not contain every illustration that appears in *Bellifortis*, but the overlap includes many of the most iconic pieces in the treatise, such as siege engines with the faces of demons, mechanical guns, diving suits, and even the spearhead inscribed with the magic word MEUFATON. The relationship between these texts is clear and unarguable. The only unanswered question is whether Vienna 3069 is based on one of Kyeser's early drafts, or whether the primevals and *Bellifortis* share a common pictorial source. GILLE declared it unanswerable in the absence of further evidence,[376] and fifty years later CERMANN was unable to be more decisive.[377]

Given that much of the other content of *Bellifortis* is copied from existing works, it seems most likely to me that the war machines were no different. Kyeser was a compiler or cataloger, not an inventor, and his role was to arrange these existing, disparate contents into a presentation that reflected his vision for modern warfare. We see his wit and intelligence in the poetic and often sardonic descriptions he added to these contents, not in the contents themselves.

Composition and Evolution

As mentioned above, Kyeser produced at least four different versions of *Bellifortis* in the first decade of the fifteenth century, and possibly another thereafter.[378] Each version rep-

---

[374] There were a variety of catalogs of war machines produced in the fourteenth century, such as the *Texaurus regis Franciae* of Guido da Vigevano, but for the most part they are demonstrably different from the art in *Bellifortis* and were at most inspirational. The primeval group is the only clear instance of direct influence.
[375] GILLE, 56.
[376] Ibid., 58.
[377] CERMANN (2013), 58-60.
[378] CERMANN (2013), 82.

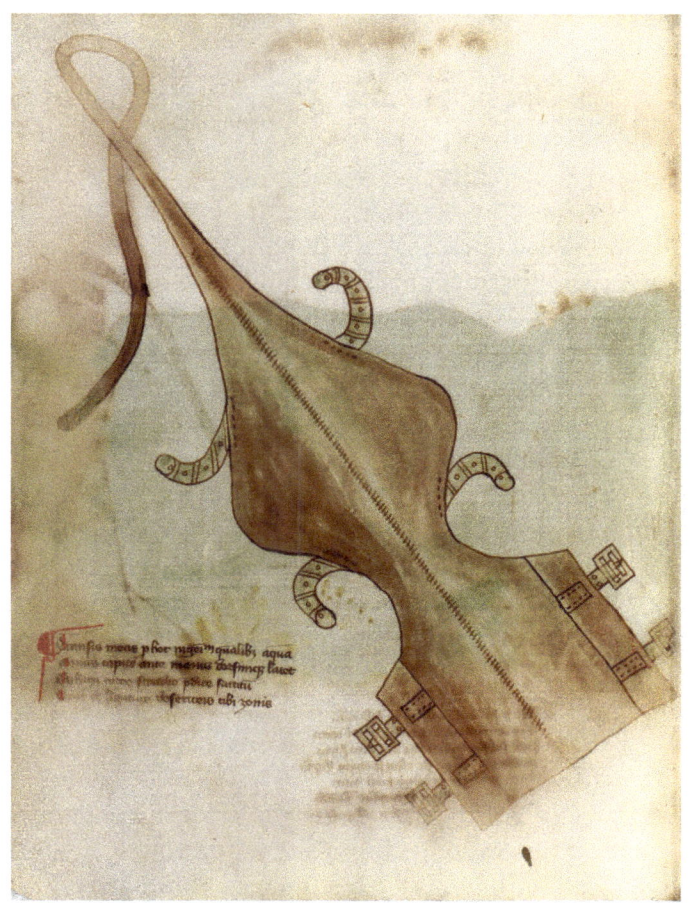

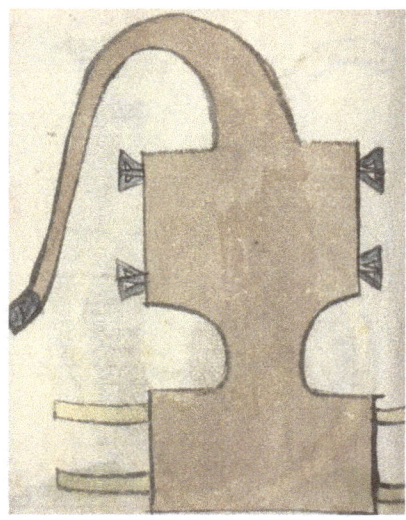

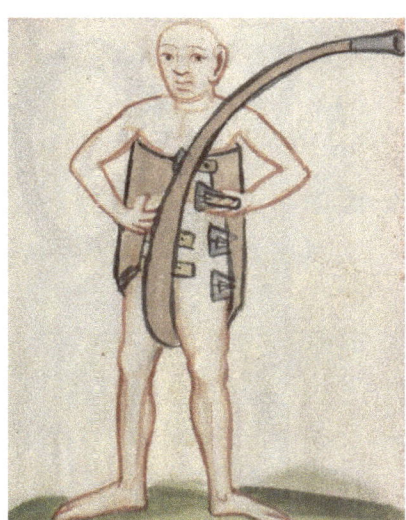

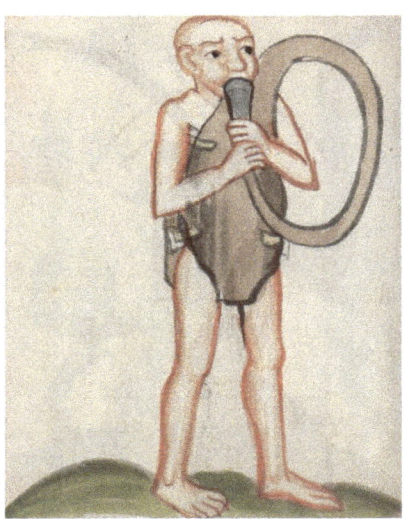

Figure 13: An inflatable life-preserver.
Legend: Göttingen 64a, fol. 50ᵛ-51ʳ;
Vienna 3069, fol. 60ᵛ, 62ʳᵛ.

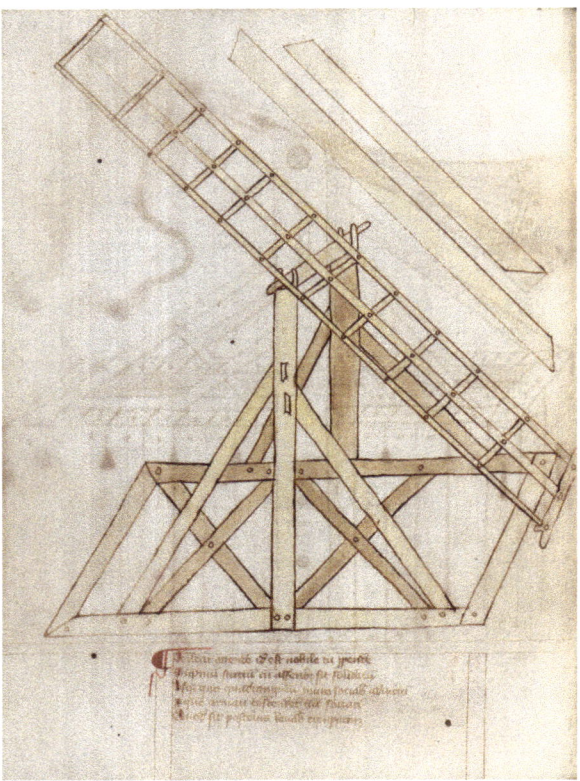

Fig. 14-15: Ladders. Göttingen 64a 21ᵛ, Vienna 3069 45ʳ.

resented a significant evolution over those that came before. These were introduced briefly in Kyeser's biography, but I will elaborate here.

The oldest known manuscript is Göttingen 64a, completed in 1402. It currently consists of 159 paper folia. Though it contains a dedication to Wenzel, it seems to be quite incomplete: about a quarter of the pages of the manuscript are blank, and many have empty frames drawn on them (indicating that content was planned for those pages but never added).

Göttingen 64a is organized into ten chapters, though not all of them received the same amount of attention. The influence of the primevals group is most apparent in this manuscript, and about half of the illustrations in 64a have an equivalent in Vienna 3069.[379] What's more, the illustrations in Göttingen 64a are often much simpler than the equivalents in Vienna 3069 (for example, 64a will display a single ladder in isolation where 3069 shows the ladder placed against a castle wall with soldiers climbing it). This makes Göttingen 64a an artistic bottleneck for *Bellifortis*: the details from the primevals group that aren't reflected in 64a do not reappear anywhere in the tradition, and elaborations by later artists diverge sharply from the primevals.[380]

The second manuscript produced by Kyeser is also owned by the Göttingen State and University Library, Göttingen 63. This was completed in 1405, and currently consists of 144 parchment folia. It was a lavish, expanded version of his first effort (or perhaps just a complete example of his interrupted plan for 64a), and boasts artwork of a much higher quality. This manuscript was initially intended to be dedicated to Wenzel, but this was changed to Ruprecht.[381]

QUARG connects the artwork in Göttingen 63 to the workshop of German illuminators

[379] A comprehensive study of all six manuscripts in the primevals group and their contributions to *Bellifortis* is lacking, and might show even more overlap in addition to suggesting a more complex history. CERMANN suggests, for example, that Weimar could have a separate origin from Vienna 3069. See CERMANN (2013), 58-60.

[380] CERMANN identifies a single point in the 1420s where a second infusion of primeval content is introduced, leading to Rome 1888 in 1430 and six later copies around the turn of the sixteenth century. The four manuscripts in the pseudo-Hartlieb branch may also be influenced by Rome 1888. See CERMANN (2013), 94-95.

[381] CERMANN (2013), 43.

that Wenzel brought to Prague.[382] The manuscript has only a few blank pages, and a significant number of new illustrations that don't come from the primevals group (suggesting the possibility of other unidentified sources). In many places, the artwork of Göttingen 63 shows new elaborations over that of 64a, but other times details are obscured by the new art style and lost.

Figure 16: An armored vehicle. Legend: Göttingen 64a, fol. 19ᵛ.

Göttingen 63 has the same ten chapters as 64a, but after this Kyeser seems to have begun rethinking his plan. His third work has been lost and only survives in two brief fragmentary copies in the Austrian National Library: Vienna 6562A and B, both created in ca. 1407. Kyeser's fourth (and possibly final) version of *Bellifortis*, however, saw the treatise reorganized into a new 7-chapter format. This manuscript is also lost, but two complete manuscript copies survive from ca. 1410, Chantilly and Rome 1994, each of which is the beginning (hyparchetype) of a large family of copies.

Figure 17: An armored vehicle. Legend: Göttingen 63, fol. 38ᵛ.

The 7-chapter format retains most of the contents of the earlier treatises, but merges some of the smaller chapters together and often moves individual devices to different chapters. (Indeed, Göttingen 64a and 63 show a fairly weak thematic organization, to the point that it seems likely that many pages were moved from their original locations.) Some additions to Göttingen 63 were discarded again in this version, and all personal references to Kyeser were omitted; henceforth he only refer to himself as the Exile.

The fifth version that Kyeser was potentially involved in was produced in the 1410s and is yet again lost, but it followed the same 7-chapter format and is exemplified by Vienna 5278 (ca. 1420). The 7-chapter edition is thus the final form of *Bellifortis*,[383] and it is the basis for

---

[382] QUARG, XVI.

[383] QUARG tried to make the opposite argument, dismissing the 7-chapter *Bellifortis* as an early draft predating Göttingen 63 which Kyeser never intended for publication. In QUARG's narrative, various drafts were found after Kyeser's early death in 1405 and copied by well-intentioned but ignorant lackeys. This is his explanation for why they proliferated and the 10-chapter did not. This seems to be another of QUARG's fantasies, and fails to conform to basic facts of chronology and the precise dates for the earliest *Bellifortis* manuscripts. Instead, CERMANN makes the argument that the end of Kyeser's involvement in the treatise should be judged by when

Vegetius	7-chapter version	10-chapter version	Elements[384]
3. Field battle	1. Field battle	1. Field battle	Earth
4. Siege war	2. Siege weapons	2. Siege weapons	
		3. Water technology	
	3. Ladders and climbing	4. Ladders and climbing	
		5. Mechanical weapons	
4. Defensive war	4. Defensive technology	6. Defensive technology	
5. Naval battle	5. Water technology		Water
		7. Flares	Fire
	6. Pyrotechnics and firearms	8. Pyrotechnics	
		9. Thermal engineering	
	7. Misc. tools and weapons	10. Misc. tools and weapons	Air

nearly all copies after 1410 (the only exceptions being Innsbruck, a copy of Göttingen 63 from the 1450s, and Istanbul, a tiny fragment copied from Innsbruck in 1480).

The basic plan for most of Kyeser's book is modeled after Vegetius, and specifically the latter two books of his treatise. Study of extant manuscripts of Vegetius show that Medieval readers tended to focus on the first one or two books, while book three received scattered attention and book four was often only read for its exotic language, not its practical advice.[385]

Kyeser may thus have intended to reintroduce his readers to a part of military strategy that he thought was wrongfully neglected: Book III discusses field battles, while Book IV discusses sieges and fortress defense; the second half of Book IV is dedicated to naval issues and sea battles, but this was sometimes separated into a Book V in Medieval versions.[386]

To see how this lines up with *Bellifortis*, see the table.[387] The structure of the 10-chapter *Bellifortis* was already somewhat similar to Vegetius, but the 7-chapter version tightens it up considerably. Kyeser's presentation of each of these subjects diverges considerably from the older authority, but the homage is clear.

Bookending this Classical framework are more esoteric materials that try to actualize Vegetius to the turn of the fifteenth century. Both versions begin chapter 1 with depictions of sev-

---

his coat of arms stops appearing. The final manuscript to display it is the Rome 1994, one of the two direct copies of the original 7-chapter manuscript. See QUARG, XXIX, and CERMANN (2013), 72-76.

[384] FRIEDRICH argues that superimposed over this order is another set of divisions based on four elements of natural philosophy. These are personified by historical-mythological figures appearing at the beginning of each segment: chapters 1-4 deal are assigned to Alexander the Great (representing the earth); then chapters 5-7 are assigned to the angel Salathiel (representing water), Philoneus (fire), and the Queen of Sheba (air), respectively. These elements are not particularly connected to the mythology of these figures, but Salathiel is depicted with jugs of water while the latter two are mechanical devices; Philoneus is a tool for lighting fires, and Sheba can blow a stream of ash onto the face of anyone who looks at her.

[385] ALLMAND, 41.

[386] ALLMAND, 44.

[387] Table inspired by a similar presentation on FRIEDRICH, 11.

en riders representing the known "planets" and describing their astrological significance.[388] This is a well-known example of the "children of the planets" genre of pop astrology from the fourteenth and fifteenth centuries.[389] Then the latter parts of the book consist of chapters devoted gunpowder, alchemy, and magic. (The final chapter in both versions displays an assortment of strange inventions and ideas that don't fit anywhere else.)

Inside of this framework, Kyeser does not pretend to repeat the teachings of Vegetius. Instead, the specter of Nicopolis appears again and again. The *ribauldequins*[390] in the field battle chapter bring to mind French knights killed through effective use of terrain and temporary obstacles, while the "chariots" (more properly "war wagons") and wagon forts show how the Ottoman cavalry could have been repelled. The siege weapons and wall-scaling equipment might have broken the fortress of Nicopolis before the Ottomans arrived, while the defensive chapter shows why the overconfident knights could never have penetrated it through force alone. The chapter on water technology offers many means of bringing armies across water obstacles, which could have saved countless lives after the crusader army began to retreat. And the use of fire (mundane and magical) along with the marvels of gunpowder provides hope to any army faced by overwhelming odds.

Most illustrations in both the 10- and early 7-chapter versions are described in Latin verses using an inconsistent dactylic hexameter (a Roman poetic form often mimicked in the Middle Ages), which may have added prestige to the treatise but certainly made it more difficult to interpret (even by educated readers of the fifteenth century) and raises questions about the intended audience of the book.[391]

Kyeser may have realized that he was limiting his influence by only catering to Latin-educated people with the patience to muddle through his poetry; CERMANN indicates that the 1407 manuscript contained a German prose translation replacing the Latin verses.[392] By the 1430s, enterprising translators had created two more German prose versions of the text, one of which

Figure 18: The battle of Nicopolis.

Legend: Jean Froissart, 15th c.; BnF ms. Français 5190 réserve, fol. 239v (detail).

---

[388] Göttingen 64a only includes the text describing the planets. The figures of horsemen were devised for Göttingen 63.

[389] For a discussion of the children of the planets, see Christian Tobler's chapter in this book.

[390] A *ribauldequin* was a type of primitive moveable gun emplacement first appearing in the fifteenth century, eventually replaced by true field artillery. The term is used by GILLE (p 60) to refer to the many similarly-shaped spiked barricades of this chapter, and no other author offers a term for them.

[391] LENG, 118, 282.

[392] CERMANN (2013), 69-70. I tried to confirm this, but the scans of Vienna 6562A and B that the library provided do not contain the text she describes, and as yet I've been unable to straighten the issue out.

Figure 19 (left): Wagons for a wagon fort. Legend: Chantilly, fol. 13ᵛ.

Figure 20 (right): Building a wagon fort. Legend: Vienna 3062, fol. 165ʳ.

is represented by the Vienna 3068 (which would then form the basis for Copenhagen, Hans Talhoffer's heavily-abridged version).

The plan of the 7-chapter *Bellifortis* was sophisticated, perhaps too sophisticated to survive through many copies. Manuscripts descending from Rome 1994 tended to have good fidelity, but those descending from Chantilly quickly fell into chaos. Out of this whole branch, only the Vienna 5278 presents the original 7-chapter organization. The other versions appearing in the 1420s and after show no coherent sequence of topics and reduce the treatise to a catalog of disparate curiosities; these are grouped together as the "disordered 7-chapter version".

At the same time that the substance of the text was fading, though, the artwork sometimes made great leaps forward. In some cases artists attempt to introduce additional details, based on either the description in the text or on their own ideas about realism. In other cases, several inventions and devices would be combined into a single scenario showing their imagined applications.

While dozens of copies and revisions of *Bellifortis* were made in the first half of the fifteenth century, interest in Kyeser's work proved to be short-lived. It was copied only sporadically in the late fifteenth century, and there are only about

Figure 21: A battle scene featuring a wagon fort and other Bellifortis *tools. Legend: Vienna 3062, 147ᵛ-148ʳ.*

five manuscripts created after 1500, with the Rome 1889 effectively ending the tradition in 1525.

Contents of the *Bellifortis*

With the plan of the treatise in mind, let us turn to a discussion of some of its specific contents. This is based on the 7-chapter *Bellifortis*, and particular attention will be paid to the pieces included by Talhoffer in the Copenhagen.

Chapter 1 begins with a parade of planetary figures followed by the representative of land war, Alexander the Great. He is accompanied by a spear inscribed with the word MEUFATON and an esoteric symbol. This is followed by another strange scene: two heavily-armored fighters face each other as if for a duel. Overhead, a smiling sun shines on them, and the text describes how a good strategy for a duel is to burnish your armor to a high shine and then place yourself facing the sun so that its light blinds your opponent. Early copies of *Bellifortis* illustrate this by drawing a ray of sunshine connecting to the fighter on the right and then shooting across at the face of the fighter on the left (who is forced to shield his eyes with one hand); this detail is lost on a lot of artists, who draw the rest of the scene but omit the sunbeam that illustrates its central point.

Figure 22: Knights in shining armor.
Legend: Göttingen 63, fol. 18ᵛ; Vienna 3068 fol. 55ᵣ.

One detail unique to Göttingen 63 is the weapons of the knights: they are wielding short daggers but holding them in the "half-sword" grip with one hand on the blade and one on the hilt. Given Kyeser's contempt for the knighthood, who he blamed for the defeat at Nicopolis, it is hard not to see this as joke at their expense—that their most effective weapon is the shininess of their armor. Later artists changed this detail as well, giving the knights more serious weapons and missing the point completely.

After this, the chapter settles down to its main theme, field battle. There are two basic forms of equipment that Kyeser presents here: *ribauldequins*[393] and war-wagons. The *ribauldequins* are a form of mobile barricade designed to be maneuvered into position by infantry and then abandoned. When used in large numbers, the device could disrupt cavalry charges with a variety of weapons on the front and sides to discourage approach and a long haft behind to brace it against the ground. These often appear fantastic and impractical in *Bellifortis*, but that's partly because the illustrations make it hard to judge their size and details of their function.

War wagons, on the other hand, are far more versatile. They were essentially heavy carts or wagons with armored walls to repel fire from bows, crossbows, and even light firearms. Some

---

[393] See note 390.

were simply intended to safely carry groups of infantry to their positions, but other more sophisticated models included ports for soldiers inside to fight back with their own crossbows or cannons. War wagons were a key strategic weapon in Eastern Europe until improvements to cannon technology made them obsolete by the late fifteenth century.

Both of these concepts are combined in the tactic that Kyeser was most impressed by, the wagon fort (or wagon castle). A group of wagons would be linked together with chains and used to create a strong barrier against both cavalry and infantry; the wagons could be arranged into many different shapes depending on the needs of the battle, and were almost impossible for an enemy force to separate. Kyeser introduces this with a picture of wagons that are essentially logs on wheels, making them somewhat similar to a row of *ribauldequins*, but the tactic gains maximum efficiency when war wagons are included in the chain (illustrated previously). This devastating tactic was key to the success of Hussite forces during the earlier stages of the Hussite Crusade of the 1420s-40s.

Chapter 1 material in Copenhagen 15ᵛ, 35ᵛ-38ᵛ.

Chapters 2 and 3, discussing siege warfare and climbing tools respectively, are not very well differentiated in extant copies.

Figure 23: A war wagon.
Legend: Vienna 3062, fol. 78ᵛ.

Figure 24: War machines.
Legend: Chantilly, fol. 20ʳ, 16ʳ.

The chapter on sieges is quite conventional in its approach: it portrays trebuchets, cranes, battering rams, and so on. The chapter on climbing presents many different designs for portable ladders. But chapter 2 also includes a few ladders, and its most interesting devices might be portable bridges for crossing moats and ramp structures for mounting castle walls. Chapter 3, after finishing with its ladders, moves on to using miners to weaken defensive walls, building screens to shield from rocks thrown down from above, an armored ram shaped like a beaked monster, and even demon summoning.

Chapter 2 material in Copenhagen: 16ᵛ-20ᵛ, 39ᵛ.

Chapter 3 material in Copenhagen: 21ʳ-24ʳ, 27ᵛ-28ʳ, 29ʳ, 40ᵛ.

Chapter 4 discusses defensive warfare, but it does so in such a cursory fashion that it's tempting to conclude that the Kyeser had no interest in the subject, and that fortress defense was only included as a formality. It devotes a few pages to the strategic positioning of fields of sharpened spikes (both wooden stakes and metal caltrops), briefly covers col-

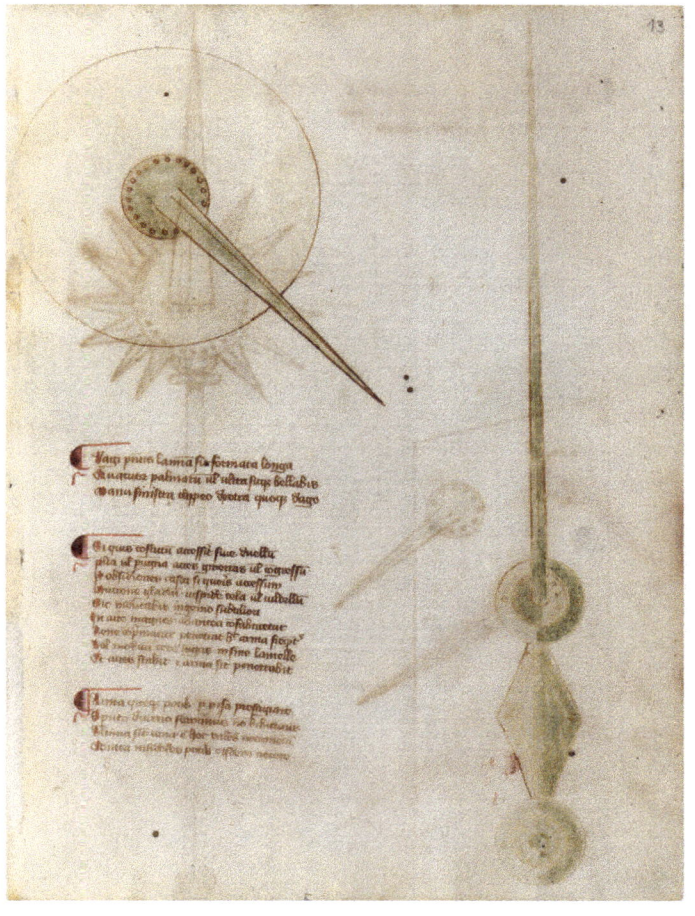

lapsible bridges, and mentions that the best guards of a fortress are dogs and geese (an allusion to the legend in which geese are credited with saving Rome).

The chapter does not end there, however, because in the 7-chapter version it has also absorbed the chapter on crossbows and other mechanical weapons. Kyeser presents no less than five different crossbow designs and a variety of special-purpose bolts. He also covers several ways of drawing crossbows, from simple leather straps to complex machines. At this point the 7-chapter *Bellifortis* diverges from the 10-chapter completely and presents several other weapons that were previously in the chapter on miscellanea. This includes two slings, two free-standing infantry shields, a spiked mace, and a rondel dagger and spiked buckler pair.

Chapter 4 material in Copenhagen: 16ʳ, 24ᵛ-25ᵛ, 29ᵛ-31ʳ, 32ʳ, 34ʳ.

Chapter 5 again shows that Kyeser wants to follow the form of Vegetius, but has no interest in the actual subject matter. Where Vegetius wrote about naval warfare, Kyeser's discussion of aquatic military technology never dips a toe in the sea. First, though, he introduces the angel Salathiel as representative of water. Salathiel appears in a few books of apocrypha,[394] and Christian tradition identifies him with the angel in Revelation 8:3-4 who carries prayers to the throne of God. There doesn't

Figure 27: A dagger and spiked buckler, and their use.
Legend: Göttingen 64a, fol. 13r; Vienna 10799, fol. 206ʳ.

Figure 25 (op. top): Mobile bridges.
Legend: Göttingen 64a, fol. 20ʳ; Chantilly, fol. 31ʳ.

Figure 26 (op. bottom): Summoning demons.
Legend: Göttingen 63, fol. 94ʳ.

---

[394] 2nd Esdras and the *Conflict of Adam and Eve.*

seem to be any association with water in any tradition I've been able to identify; nevertheless, here he stands pouring jugs of water on the ground to signify the change in subject.

Kyeser has a lot of ideas about water, and they don't form a cohesive theme. First come six different methods of crossing bodies of water, ranging in complexity from a guide cable that runs across the water, to a barge with wheels that converts into a bridge. Then he moves to personal aquatic equipment, showing three different inflatable life preservers and two full diving suits, complete with breathing apparatus. (Later copyists add a third diver, perhaps thinking that the two illustrations were trying to show too many details at once.)

From here, he pivots to other sorts of water-based technology which have even less connection to warfare. He portrays rough versions of Archimedes' screw, Hero's fountain, a water wheel, a chain pump, siphon-based systems for transporting water uphill, and a hand-cranked paddle boat (the closest he comes to actual maritime matters). He also presents snowshoes and two different bathhouses. All in all, a strange combination, but that will characterize most of the remainder of the book.[395]

Chapter 5 material in Copenhagen: 26ʳ, 27ʳ, 31ᵛ, 35ʳ, 40ʳ, 41ʳ, 43ᵛ-44ʳ, 45ʳ, 46ᵛ.

Chapter 6 covers the broad subject of fire and pyrotechnics, combined from three separate chapters from the 10-chapter version. This is where the text of *Liber ignium* appears, as well as selections from the *Experimenta Alberti* and *De mirabilibus mundi*. It also presents several complex cannon arrangements of a type that historians call "organ guns",[396] one of the earliest depictions of a culverin,[397] and wooden wagons designed to shelter gunners (similar to those in Chapter 1). Kyeser's other fire-based inventions

---

[395] Note that an error was introduced in this chapter early in the tradition. The guide rope for horses on Chantilly folio 101v and the swimming harness on folio 102r have their captions and illustrations swapped in both in Vienna 5278 and the Vienna 3068; these two manuscripts aren't directly connected, suggesting that this error arose during the 1410s. This becomes particularly problematic in Copenhagen (Talhoffer), which only includes one of the two devices—the illustration of the diving belt and the caption for the guide rope appear on 26r (which sounds like nonsense in that context), and the other caption and illustration are omitted entirely.

[396] Because the arrangement of the gun barrels had the appearance of a pipe organ.

[397] A small cannon that could be carried and fired by a single person (with difficulty).

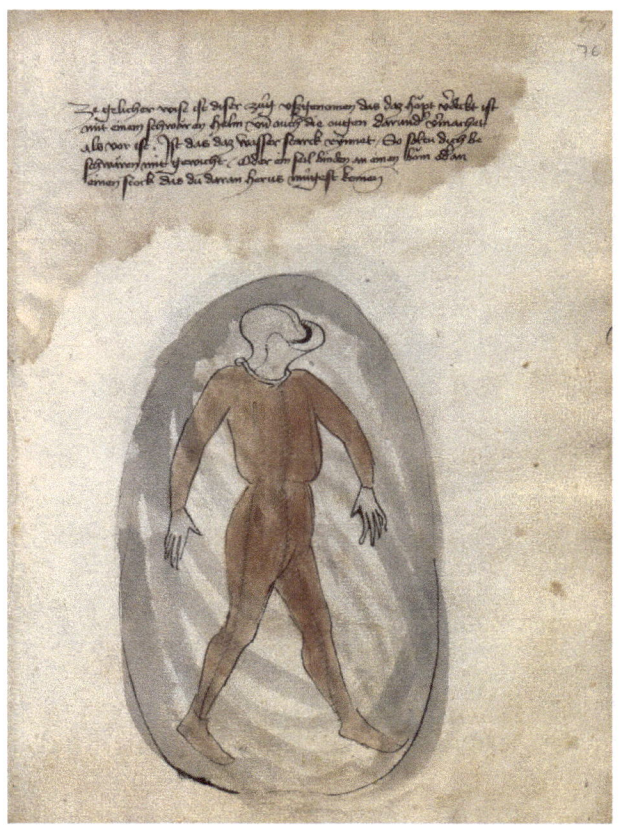

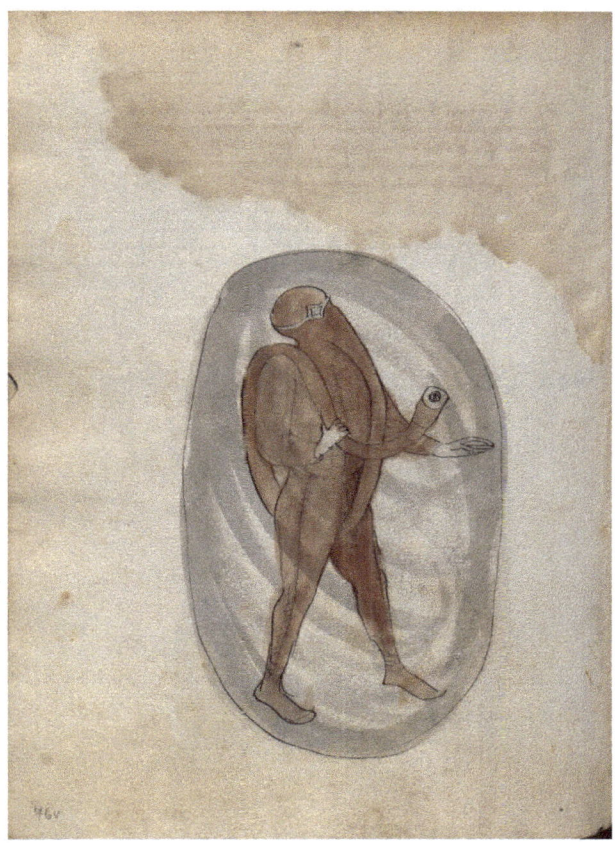

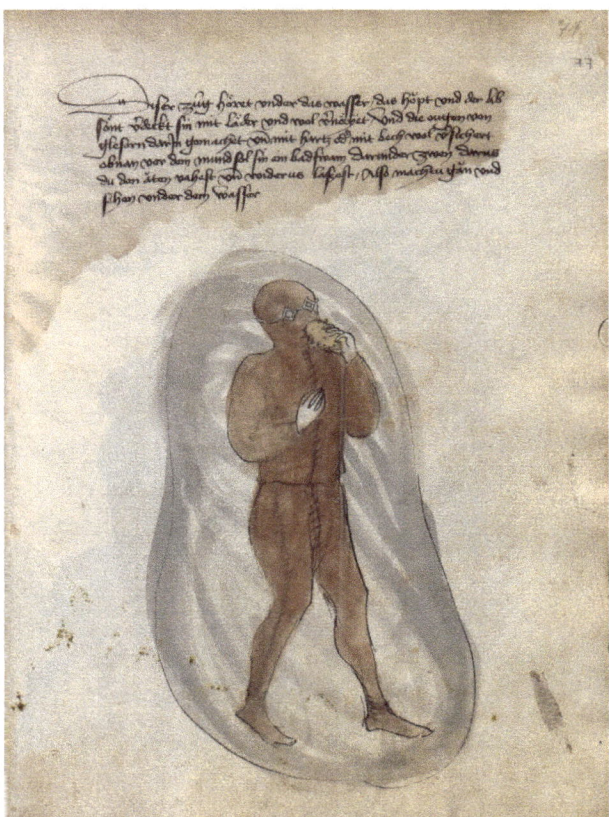

Figure 26 (op. top): Diving suits.
Legend: Göttingen 63, fol.62ʳ.
Figure 27 (op. bottom): A culverin.
Legend: Göttingen 63, fol. 104ᵛ.

Figure 28: Four diving tools.
Legend: Vienna 3068, fol. 76ʳ-77ᵛ.

Figure 29: A dragon balloon.
Legend: Göttingen 64a, fol. 131ᵛ.

Figure 30 (op.): Philomenus in his original and misunderstood forms.
Legend: Göttingen 64a, fol.130ᵛ; Vienna 3068, fol.79ᵛ.

Figure 31 (op.): Queen of Sheba in her original and more developed forms.
Legend: Göttingen 63, fol. 122r; Frankfurt 184ᵣ.

include rockets, bombs, kilns, and a dragon-shaped aerostat[398] or balloon.

Chapter 6 is presided over by a figure called Philoneus (or Philomenus), representing fire. This appears to be a classical reference, but the historical Philoneus that Kyeser was referencing has yet to be identified.[399] In *Bellifortis*, his name is applied to a metal statue containing a fuel reservoir and is designed to spout fire from the mouth when heated. Early artists seem to have misunderstood this illustration, because his metallic skin disappears early in the tradition and he is drawn as a nude man; subsequent readers might have thus been confused by the description of filling him with fuel so he could breathe fire.

Chapter 6 material in Copenhagen: 32ᵛ, 34ᵛ, 39ʳ, 41ᵛ-43ʳ, 45ᵛ-46ʳ.

Chapter 7 has no clear unifying theme. It is described as "various natural weapons", which was accurate in the 10-chapter version, but most of those weapons are moved elsewhere in the 7-chapter version. What remains is a strange catalog of devices: a bellows-operated air mattress, two different wind-powered elevators, a goose weighed down by an anchor, several more magical and herbal recipes, a comical "Florentine apron",[400] a collection of small hand tools, a male castration device, and what seem to be extra *ribauldequin* and war wagon designs that didn't make it into chapter 1.

While some of these devices have practical value, the 10-chapter version states that some of the contents of this chapter should be read as humorous (a line that was left out of the 7-chapter version). An example might be the presentation of the Queen of Sheba who presides over the final element, air. She is drawn with black skin in the Göttingen 63, but this detail is quickly lost and generally she is portrayed as white woman. On her chest is a circular mirror, and the text states that when young men come to stare at her chest, she can exhale a cloud of soot from a hidden device, leaving their faces black as hers.

---

[398] Gille, 64.

[399] There are various well-known Classical figures with similar names, but none has an association to fire.

[400] This is commonly considered the earliest depiction of a chastity belt, and is entirely fictional in that it reflects no real object or practice. Cermann traces it to a satirical novel about jealousy by Giovanni Sercambi, written a few years earlier in the late fourteenth century and probably introduced to Kyeser during his stay in Padua. There's no evidence that such a device was ever produced in this time period, but the joke took on a life of its own and in the fifteenth century it was attributed to the long-deceased Francesco I da Carrera. From there it passed into popular lore and especially captured the imaginations of writers from the 19th century to the present, who fixate on the practical implications of such a device and imagine a variety of motivations about why such a device would be invented. Such speculation, Cermann observes wryly, says more about modern writers than it does about the device or its time period. For her complete findings, see Cermann (2013), 67-68.

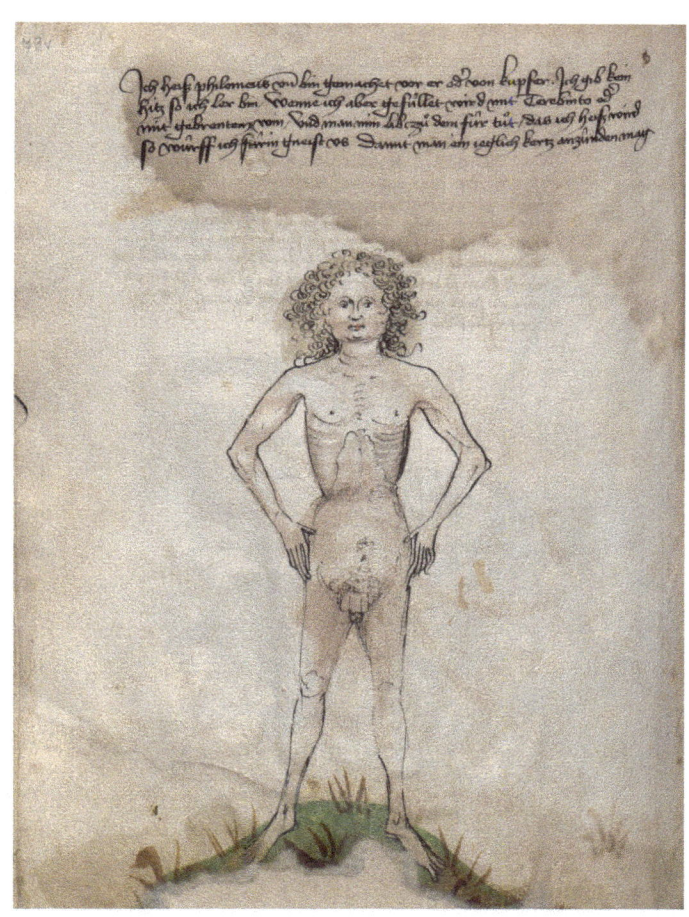

Figure 32: A complex scene.
Legend: Copenhagen, fol. 12ᵛ.

The fact that this text accompanies a figure of a white woman in even the earliest copies of the 7-chapter version is more evidence that the artists didn't necessarily understand the captions. Other artists clearly understood more, adding mechanical elaboration to illustrate the mechanism behind it (even as they painted the wrong skin color).

Chapter 7 material in Copenhagen: 22ʳ, 25ᵛ, 28ᵛ, 33ʳᵛ, 44ᵛ, 47ᵛ.

Over time and the course of many copyings, the 7-chapter *Bellifortis* acquired additional material (such as the extra diving suits mentioned above). The first such addition, preserved on folia 162ᵛ-173ᵛ of Vienna 5278,[401] consists of 22 mostly-uncaptioned illustrations that may been devised by Kyeser himself. About half of these are new ladders and climbing tools, perhaps intended to bolster the lackluster chapter 3 in a future revision that never occurred. The rest include complex arrangements of pulleys and winches, a new *ribauldequin* and war wagon, a crane operated by horses, and a windmill.

Other changes to *Bellifortis* are compressions rather than expansions: some artists attempt to combine several individual

---

[401] Four of these devices also appear in the Vienna 3068; as previously stated, these two manuscripts aren't directly connected, suggesting that the devices were added in a lost parent copy from the 1410s; one device from Vienna 3068 was then copied into Copenhagen. Vienna 3068 adds a new recipe for bread to be eaten during a siege, which is also passed to Copenhagen.

Figure 33: A complex scene.
Legend: Copenhagen, fol. 14ʳ.

devices from the text into a more elaborate scenario. One such scene has already appeared above, Vienna 3062's depiction of a wagon fort being used on a battlefield along with large infantry shields and cannons. Copenhagen begins its presentation of *Bellifortis* with six composite scenes.[402] For example, 14ʳ shows a man in water clinging to a rope span (chapter 5), while a man on the shore holds a crossbow in the process of some sort of rescue attempt using one of the special crossbow bolts (chapter 4). A second man on the shore holds a knife and a pot, but it's unclear what he is doing with them. 12ᵛ shows another crossbowman firing at targets (chapter 4), a large firebomb (chapter 6), an air mattress (chapter 7), and a peacock.

Conclusion

We know much about Kyeser's times, but almost nothing about his own life. Outside of a single record indicating that a person named "Conradus Kyeser" was alive and in Prague in 1390, he seems to have eluded the gaze of history entirely. We have found no records of a Kyeser family or Konrad's coat of arms in fourteenth century Eichstätt, nor have we found a Konrad Kyeser in the service of any of the princes he lists. This is a

---

[402] Copenhagen, 12r-14v.

rather surprising absence for someone who boasts of his international renown.[403]

Likewise, his manuscripts do not attempt anything resembling an autobiography, even though the few personal details we can scrape from his writings tend to be presented that way. Information about his birth and parents comes from a personal horoscope that he included in the work. His relationships to princes and nations are reconstructed from boasts about his renown or personal attacks against them. Even his participation in the great Battle of Nicopolis is a later addition, not mentioned in his 1402 work but gaining a prominent position in 1405.

I would like to propose an alternative hypothesis: we cannot locate Konrad Kyeser in history because he never existed (as we know him).

Around the turn of the fourteenth century, another Konrad earned a medical degree and began to practice medicine in Eichstätt. He was a prosperous businessman and a friend of the Benedictine abbot at Sankt Emmeram. Later in his life, he authored at least two collections of medical remedies based on Arabic teachings. By the turn of the fifteenth century, a copy of these writings[404] had made its way into the library of Wenzel IV.[405]

Perhaps a courtier in Wenzel's court, having completed a collection of inventions and occult recipes, chose to take a pseudonym in the finest Medieval tradition, settling on Konrad of Eichstätt. When his master was captured by Sigismund, his plans necessarily shifted. He became Konrad "the judge" (*Kieser*), ready to condemn the King of Hungary for his treachery. He continued this posture in 1405, but after it failed to gain any traction for his treatise, he abandoned the pseudonym and satisfied himself with being merely "the Exile" in his subsequent revisions.

Regardless of Kyeser's status, though, *Bellifortis* is an important text. Its importance doesn't lie in its novelty or inventiveness, because it is seated firmly in the larger body of Medieval military literature. It also doesn't lie in its influence over the military culture of the fifteenth century, because by all indications it had little or none.

Rather, by collecting many different strains of military (and non-military) writings into a single treatise, coherent or not, it provides us with a window on a part of the past that is poorly understood. Like Vegetius before him, Kyeser reproduces, abridges, or summarizes teachings that were common in his time but lost to us, and in so doing he preserves at least the echo of them for our benefit.

---

[403] Göttingen 63, 137r.
[404] Munich 321.
[405] LENG, 113.

Bibliography

Primary sources (with abbreviations)

Besançon	Bibliothèque municipale, Ms. 1360
Budapest	Bibliothek der ungarischen Akademie der Wissenschaften, K 465
Chantilly	Musée Condé, Ms. 348
Colmar	Bibliothèque municipale, Ms. 491
Erlangen	Universitätsbibliothek, B 26
Frankfurt	Universitätsbibliothek, Ms. germ. qu. 15
Gotha	Universitäts- und Forschungsbibliothek Erfurt/Gotha, Chart. A. 558
Göttingen 63	Niedersächsische Staats- und Universitätsbibliothek, 2° Cod. Ms. philos. 63
Göttingen 64	Niedersächsische Staats- und Universitätsbibliothek, 2° Cod. Ms. philos. 64
Göttingen 64a	Niedersächsische Staats- und Universitätsbibliothek, 4° Cod. Ms. philos. 64a
Heidelberg	Universitätsbibliothek, Cod. Pal. germ. 787
Innsbruck	Tiroler Landesmuseum Ferdinandeum, FB 32009
Istanbul	Topkapi Sarayi Müzesi, Cod. 77
Köln	Historisches Archiv der Stadt, Best. 7020 [W*]232
Copenhagen	Kongelige Bibliotek, Thott 290 2°
Munich 321	Bayerische Staatsbibliothek, Clm 321
Munich 356	Bayerische Staatsbibliothek, Cgm 356
Munich 30150	Bayerische Staatsbibliothek, Clm 30150
New York 58	Public Library, Spencer Collection, Ms. 58 93
New York 104	Public Library, Spencer Collection, Ms. 104
Paris	Bibliothèque nationale de France, Ms. Lat. 17873
Rome 1888	Biblioteca Apostolica Vaticana, CPL 1888
Rome 1889	Biblioteca Apostolica Vaticana, CPL 1889
Rome 1986	Biblioteca Apostolica Vaticana, CPL 1986
Rome 1994	Biblioteca Apostolica Vaticana, CPL 1994
Weimar	Herzogin Anna Amalia Bibliothek, Cod. Fol. 328
Vienna 3062	Österreichische Nationalbibliothek, Cod. 3062
Vienna 3068	Österreichische Nationalbibliothek, Cod. 3068
Vienna 3069	Österreichische Nationalbibliothek, Cod. 3069
Vienna 5278	Österreichische Nationalbibliothek, Cod. 5278
Vienna 5342	Kunsthistorisches Museum, KK 5342
Vienna 5518	Österreichische Nationalbibliothek, Cod. 5518
Vienna 6562A	Kunsthistorisches Museum, KK 6562A
Vienna 6562B	Kunsthistorisches Museum, KK 6562B
Vienna 10799	Österreichische Nationalbibliothek, Cod. 10799
Wolfenbüttel	Herzog August Bibliothek, Cod. Guelf. 161 Blank.

Secondary sources

ALLMAND

ALLMAN, CHRISTOPHER (2011). *The De Re Militari of Vegetius: The Reception, Transmission and Legacy of a Roman Text in the Middle Ages.* Cambridge: Cambridge University Press.

CERMANN (2013)

CERMANN, REGINA (2013). "Der ‚Bellifortis' des Konrad Kyeser". *Codices Manuscripti & Impressi,* Supplementum **8**. Hollinek.

CERMANN (2014)

> CERMANN, REGINA (2014). "Von nachlässigen Schreibern und verständigen Buchmalern. Zum Zusammenspiel von Text und Bild in Konrad Kyesers ‚Bellifortis'". Ed. CHRISTINE BEIER and EVELYN KUBINA. *Wege zum illuminierten Buch*: 245–270. Böhlau publishing house.

CLARKE

> CLARKE, STUART (2010). *Medieval Fight Book* [Television]. Wild Dream Films.

EAMON

> EAMON, WILLIAM (1983). "Technology as Magic in the Late Middle Ages and the Renaissance". *Janus* **70**: 171-212.

FRIEDRICH

> FRIEDRICH, UDO (1995). *Konrad Kyeser, Bellifortis. Feuerwerkbuch. Farbmikrofiche-Edition der Bilderhandschriften Göttingen, Niedersächsische Staats- und Universitätsbibliothek, 2° Cod. Ms. philos. 64 und 64a Cim.* Trans. FIDEL RÄDLE. Munich: Helga Lengenfelder.

GILLE

> GILLE, BERTRAND (1966). *Engineers of the Renaissance*. London: Lund Humphries.

HALL

> HALL, BERT S (1979). *The Technological Illustrations of the So-called 'Anonymous of the Hussite Wars': Codex Latinus Monacensis 197*. Wiesbaden: Reichert.

KIECKHEFER (1997)

> KIECKHEFER, RICHARD (1997). *Forbidden Rites: A Necromancer's Manual of the Fifteenth Century*. University Park: Pennsylvania State University Press.

KIECKHEFER (2000)

> KIECKHEFER, RICHARD (2000). *Magic in the Middle Ages.* Second Edition. Cambridge: Cambridge University Press.

LÁNG

> LÁNG, BENEDEK (2008). *Unlocked Books: Manuscripts of Learned Magic in the Medieval Libraries of Central Europe.* University Park: Pennsylvania State University Press.

LENG

> LENG, RAINER (2002). *Ars belli. Deutsche taktische und kriegstechnische Bilderhandschriften und Traktate im 15. und 16. Jahrhundert.* Weisbaden: Reichert Verlag.

QUARG

> QUARG, GÖTZ (1967). *Conrad Kyeser aus Eichstätt: Bellifortis: Umschrift und Übersetzung von Götz QUARG.* Düsseldorf: VDI-Verlag GmbH.

PRICE

> PRICE, BRIAN R. (2016). "Conrad Kyeser and His War-Book *Bellifortis.*" *Medieval Warfare* **6**(5): 46-51.

WHITE

> WHITE, LYNN JR. (1969). "Kyeser's 'Bellifortis': The First Technological Treatise of the Fifteenth Century". *Technology and Culture* **10**(3): 436-441.

Appendix: *Bellifortis* in fencing manuals

The most famous copy of *Bellifortis* in a fencing manuscript is, of course, the Copenhagen, which has even been featured in TV documentaries.[406] This version is often mistakenly ascribed to Hans Talhoffer in mainstream literature even though its derivation from the work of Kyeser is clear.

It is a very incomplete copy of Vienna 3068, which contains a German prose translation and is itself a disorganized copy derived from the same lost manuscript as Vienna 5278 and Rome 1488. Talhoffer's manuscript also contains a series of six unique landscape paintings (ff 12^{r}-14^{v}) which construct some of the simpler items in *Bellifortis* into more complex scenes demonstrating their use.

Among the fencing manual corpus, *Bellifortis* is connected not only with Talhoffer but also with the so-called "Blume des Kampfes" group. Apart from Copenhagen, there are three fencing manuscripts containing some version of *Bellifortis*.

- Vienna 5278 was created in the 1420s and contains the oldest known copy of "die Blume des Kampfes". It includes a complete copy of the Latin 7-chapter *Bellifortis*, and seems to be a faithful copy of the lost manuscript that translated to German to produce Vienna 3068.

- Wolfenbüttel, owned by the Herzog August Library, was created in the 1460s or 70s and contains a fairly complete copy of *Gladiatoria* as well as large number of illustrations connected in some way to the "Blume des Kampfes" tradition. The last 30 folia contain fragment of *Bellifortis* whose German translation doesn't match any other known copies; it's impossible to say how it fits into the big picture of the tradition at this time. Alternatively, it might derive from the primevals group.

- Erlangen, owned by the Erlangen-Nuremberg University Library, was created between 1490 and 1500 at the behest of the famous knight and commander Ludwig VI von Eyb zum Hartenstein. It is a complex work, including the only version of "die Blume des Kampfes" with captions for its illustrations as well as a variety of gunmaster treatises, magical texts, and a complete copy of *Bellifortis*. Its *Bellifortis* seems to combine portions of the German translation in Rome 1888 as well as Weimar (another member of the primevals group).

There are three more copies of *Bellifortis* which might be noted as being connected to the fencing genre, though not true fencing treatises.

---

[406] For example, CLARKE.

- Rome 1488, owned by the Vatican Apostolic Library, was created in ca. 1430 and contains an interesting series of drawings on 77^v-91^v which portray fencing, wrestling and jousting, including a very rare depiction of a duel between a man and a woman. The copy of *Bellifortis* in this manuscript is a German prose translation from the same lost manuscript as Vienna 5278 and Vienna 3068 (which includes an entirely different German translation).
- Vienna 3062, owned by the Austrian National Library, was created in 1437 and is an early entry in the "Pseudo-Hartlieb" branch of the *Bellifortis* tradition (which is mostly beyond the scope of this article). It contains three drawings of different duel formats (a lethal duel on horseback, a nonlethal duel on horseback, and a duel on foot with swords and longshields), as well as the *Liber ignium* of Marcus Graecus and the *Onomantia* of Johannes Hartlieb, both of which appear in other fencing manuals. Its *Bellifortis* copy is notable for expanding some of the simpler artwork into more complex battle scenes.
- Vienna 6562A and B, also owned by the Austrian National Library, are two short manuscript fragments created between 1405 and 1410, copied from Kyeser's third version of *Bellifortis*. These fragments were once bound together with another manuscript by Hans Talhoffer, Vienna 5342 (created in the 1480s), but have since been separated.

And finally, two more fencing treatises contain brief fragments of war books without being proper copies of *Bellifortis* at all (and as such, are not listed on the stemma):

- Gotha, owned by the University and Research Library Erfurt/Gotha, was completed in ca. 1448 and is yet another manuscript once owned by Hans Talhoffer, though it's not clear whether he designed it or merely purchased it already complete. In either case, it can be seen as a rough model for the Thott manuscript, though inferior in almost every way. 141^r-148^v contain sixteen rough drawings of ladders, tools, and a diving suit, all with short captions. This fragment is very similar to *Bellifortis*, but doesn't match any known copy and might represent another derivation of the primevals group.
- Vienna 10799, owned by the Austrian National Library, was completed in 1623. It is the youngest copy of "die Blume des Kampfes" and contains hundreds of fine watercolor illustrations without captions; most are copied from Erlangen, some are similar to Vienna 5278 (though evidence of direct copying is lacking), and a few are entirely unique (or copied from an unknown third source). In copying Erlangen, the creators didn't distinguish between the different sections and two illustrations from *Bellifortis*

depicting armored fighters were copied along with the fencing plays. Another armored fighting scene was devised for the spiked buckler and long rondel dagger.

To show how these manuscripts connect with the original copies authored by Kyeser, I have devised this limited stemma:

Figure 34: Partial stemma of the Bellifortis tradition.

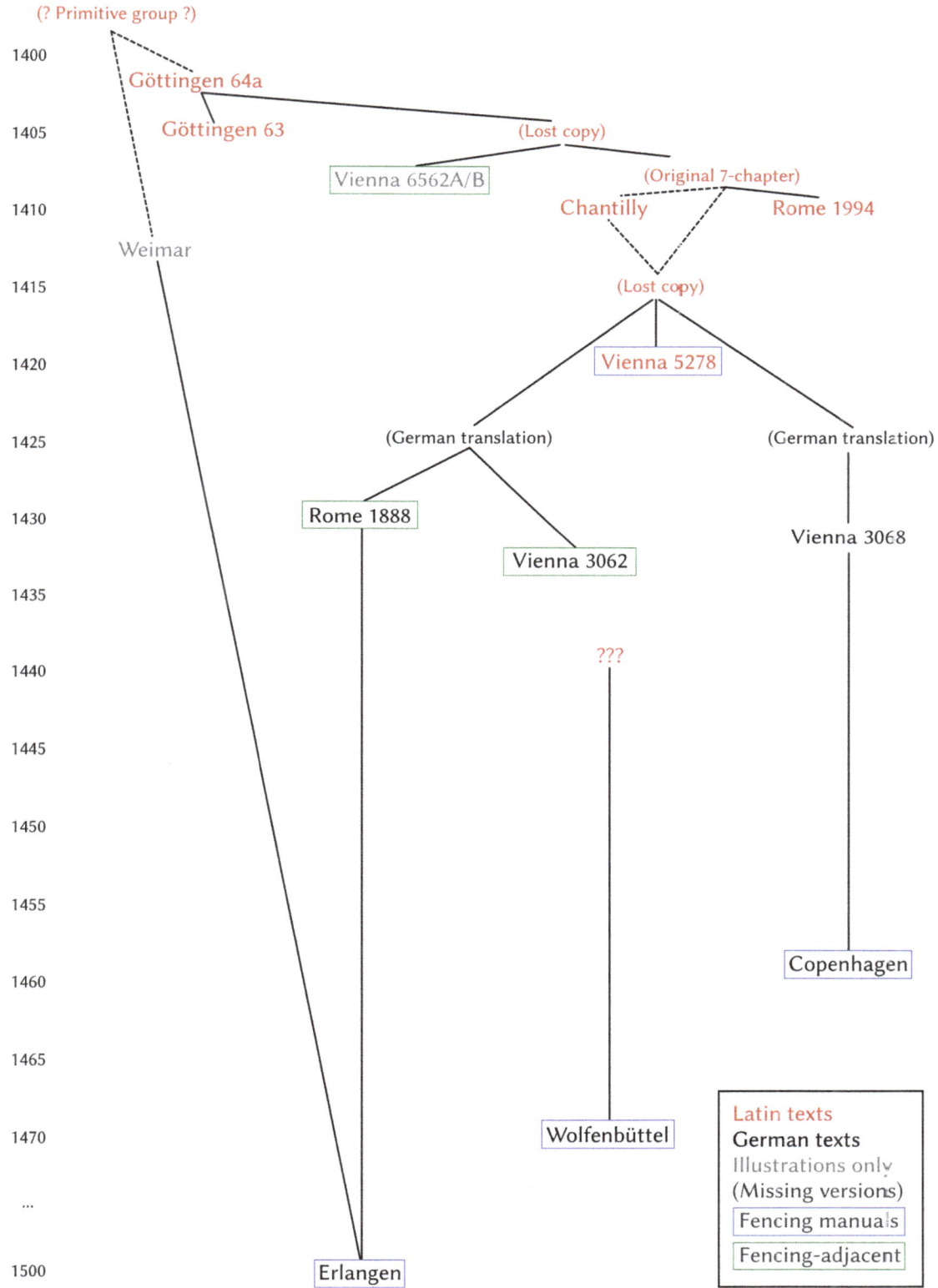

5. Tradition, Innovation, Re-enactment: Hans Talhoffer's Unusual Weapons

Ariella Elema

The fifteenth-century manuscripts attributed to Hans Talhoffer contain some of the most easily recognizable and most reproduced images in the European fight book corpus. In particular, Talhoffer's sequences of drawings depicting judicial duellists on foot battling with tall shields, and a woman fighting a man, spark curiosity both inside the historical European martial arts community and outside it: the unusual clubs, the fancifully shaped shields taller than their users, and the tight body suits seem utterly unlike most other images of combat produced in his period or before it.[407] Where did such equipment come from? Was it ever employed in a recorded judicial duel? Are there examples to be found beyond Talhoffer and the lineage of German *Fechtbücher* that followed his example?

Figure 1: Duel with swords and spiked shields.
Legend: Anonymous, 1437. Vienna, Österreichische Nationalbibliothek, Cod. 3062, fol. 28ʳ.

Historiography

The literature on the history of trial procedure in German-speaking lands has remarkably little to say about duellists' weapons and equipment in trials by battle. While Jacob Grimm briefly discussed the existence of judicial duels as a means of legal proof, he was primarily interested in the way that they contributed to the development of rules of evidence, and not in their literal enactment. Friedrich Unger, the first to write an entire book about German trials by battle, was preoccupied with their supposed Germanic origins. Although he referred in passing to weapons several times, he did not seem to think their form needed to be elaborated upon. Of the nineteenth-century history-

---

[407] The manuscripts of Talhoffer containing the sequences with long shields are the following: Gotha, Universitäts- und Forschungsbibliothek, MS Chart.A.558, ff. 24v-48r; Copenhagen, Kongelike Bibliotek, MS Thott 290.2, ff. 97v-110r; Berlin, Stiftung Preußischer Kulturbesitz MS 78.A.15; ff. 37v and 62r-77v; and Munich, Bayerische Staatsbibliothek, Cod. icon. 394a, ff. 53v-86r. Two of these manuscripts contain a duel between a man and a woman: MS Thott 290.2, ff. 80r-84r, and Munich Cod. Icon. 394a, ff. 122v-126v. A fifth manuscript, belonging to the private collection of the Königsegg-Aulendorff family, does not contain these sequences, and two later copies, in the Kunsthistorisches Museum in Vienna and the Universitätsbibliothek Augsburg respectively, are also without them. Digitizations of all the manuscripts can currently be viewed at the website Wiktenauer at https://wiktenauer.com/wiki/Hans_Talhoffer.

Figure 2: Fighting with clubs and shields.
Legend: Ethica, Aristotle, 1374.
Brussels, Koninklijke Biblio-theek van België, ms.
11201-02, fol. 341ʳ.

ians, only Henry Charles Lea and Joseph Würdinger contemplated the subject, finding a handful of references in medieval German legal codes.[408]

In the early twentieth century, Hans Fehr knew that judicial duels had sometimes been fought with weapons other than swords but elided this fact, first discussing humankind's progression from wrestling, to fist fighting, to the use of maces, then insisting that his book was not about those things. Only in his endnotes does he observe that some medieval German regions did, indeed, hold judicial duels with clubs. (We may perhaps suspect Fehr, a junior professor and former corps student when he wrote the book in 1908, of wanting to protect duelling culture's aristocratic image at a time when lethal duels were still very much alive in the German academy but increasingly condemned outside it.) Hermann Nottarp stands virtually alone among the legal historians for discussing weapons at any length. However, concerning the origins of tall shields and faceted clubs, the evidence provided by all of these scholars circles back to the manuscripts of Talhoffer and the *Fechtbuch* lineage that followed him.[409]

More recent books, such as those of Bartlett and Neumann, concern themselves with the *mentalités* that powered the judicial process. This focus is partly a consequence of the surviving sources. Courts of law in German speaking lands adopted written recordkeeping centuries later than their English and French counterparts. There are very few charters or case records from which to draw details of individual duels, and information must be gleaned mainly from law books and chronicles. In addition, judicial duelling was always an unusual way to resolve a lawsuit. As Stephen White has noted, medieval litigants in trials by battle usually managed to broker a settlement outside of court before they arrived at the combat stage of the process.[410]

On the other side of the coin, martial arts research is very much aware of the German *Fechtbücher* but, with the excep-

---

[408] Grimm, Deutsche Rechtsalterthümer, 2nd ed., 929. Unger, Der gerichtliche Zweikampf, 11, 22, 55. Würdinger, Beiträge, 13, citing Paulus Kal's Fechtbuch. Lea, Duel and the Oath, 154, citing Würdinger and reproducing an image from the duel between a man and a woman.

[409] Fehr, Der Zweikampf, 4; 41, n. 7; 46, n. 37. Nottarp, Gottesurteilstudien, 283-7.

[410] Bartlett, Trial by Fire and Water, 103-126. Neumann, Der gerichtliche Zweikampf. Two neglected collections of case law for late medieval German trials by battle can be found in Jung's Miscellaneorum, pp. 182-213, and Zwierlein's Abhandlung, no. 16, pp. 31-6. On brokering settlements: Stephen D. White, "Proposing the Ordeal and Avoiding It," 89-123.

tion of Fortner, gives almost no notice to the great volume of legal sources that preceded them, struggling to find examples of clubs or maces, shields, and clothing analogous to the ones Talhoffer illustrated. A few scholars of archaeology see similarities between surviving weapons and the Talhoffer images. Fărcaş notes the resemblance between the "prismatic" bronze mace heads found in Transylvania and the faceted wooden clubs of the German *Fechtbücher*, but he does not suggest that the design was transmitted from one place to the other. Talaga sees a relationship between the faceted pommel of an estoc in Wawel Castle in Kraków, the spiked pommels of the weapons depicted in the Thott manuscript, and the treatise of Filippo Vadi. He concludes on this basis that the estoc was a judicial combat weapon, without providing evidence for judicial duelling with edged weapons in Poland in the late fourteenth or early fifteenth centuries, the period to which the sword is dated.[411]

The missing piece in all these works is an overview of the evidence for specialized judicial duelling equipment and practice in the medieval German realm that Talhoffer knew. This article attempts to fill that gap.

Clubs and Maces

In most of Europe, for most of the history of trial by battle, the usual weapon of judicial combat was not a sword but rather a wooden club (see fig. 2). It had not always been thus: the so-called barbarian laws of the early Middle Ages do indeed mention duels with swords, but these would later fall out of fashion. The laws of the Ripuarian Franks, originating around Cologne circa 630 CE, speak of judicial combat "*cum gladio*" in disputes about whether a freedman had really been freed, while the laws of the Alemanni farther south, compiled around 730, also record a specific legal procedure by means of "*spata tracta*". Moving west, the early ninth century bishop Agobard of Lyons railed against the laws of the Burgundians, which allowed duels with spears and swords (*telis et gladiis*).[412]

At the beginning of the ninth century, a change seems to have occurred throughout the Carolingian Empire, which at the time encompassed most of the land that makes up modern France, Germany and northern Italy. The emperor Charlemagne issued a capitulary in 803 creating an addendum to the laws of the Ripuarian Franks and providing several alternatives for settling a lawsuit about whether one man had wounded another. One of the options was to hold a judicial duel, but this time the combat

---

[411] Fortner, "'Kempflich angesprochen:' über Kampfgerichte und Kampfrecht," pp. 10-22. Fărcaş, "Maces in Medieval Transylvania," 76-84. Talaga, "A Kampfschwert from the 15th Century," 8.

[412] Lex Ribvaria, ed. Beyerle & Buchner, § 60.2, p. 108. Lex Alamannorum, ed. Lehman & Eckhardt, 2nd ed., § LIV.2 (LVI.1), p. 118. Agobard, Adversus legem Gundobadi in Agobardi Lvgdvnensis: Opera omnia, ed. L. Van Acker, c. XIII, p. 27.

was to be fought not with swords but with a shield and club (*scuto et fuste*). In 818 or 819, around the same time that Agobard was writing his treatise, Charlemagne's heir Louis the Pious extended the procedure to cases of theft, in a capitulary that spread more widely across the empire.[413] From that point forward, clubs became the normal weapons of judicial combat in most of the Frankish world and swords became the exception.

German laws of the high and late Middle Ages need to be seen in the context of this widespread use of wooden clubs for judicial duelling. In the French-speaking lands that had once been the western half of the empire, legal customals and the lawsuits recorded in charters speak of duels with clubs in all instances except criminal cases where the accused was a nobleman. In all other criminal cases the prescribed weapon was a wooden club, and in all suits concerning the possession of land the battle, if there was one, was fought by champions with clubs. "Champions should not fight each other with clubs that exceed three feet in length to the base of the hand," wrote King Philip Augustus in 1215, "or they can fight with a shorter length, if they want." When such weapons are depicted in high medieval illustrations, they resemble short baseball bats. In England and some jurisdictions on the Continent, duelling clubs had "horned" or T-shaped heads, ranging in shape from modest knobs to spikes resembling pick-axes. This latter type were referred to in Latin as *baculi cornuti* or in French as *bastons cornus*.[414]

French legal practice with regard to duelling weapons influenced England, where trial by battle was introduced by the Normans after 1066, but also the French-speaking parts of Flanders and sections of northern Iberia. The *Fuero general* of Navarre from 1238 stipulates that suspected thieves should fight judicial duels "*á escudo et á baston*".[415]

Moreover, the customary use of clubs extended even farther afield. Capitularies of Charlemagne and his grandson Lothair were recorded in the eleventh-century *Liber Papiensis* compiled in Pavia, and thereby became the basis of the Lombard law that formed a foundation of precedents for the patchwork of regional secular law throughout northern Italy. "Where it is clearly apparent that he who initiated an accusation or he who is defending wishes to perjure himself, it is better that they go into the field and fight fairly with clubs than that they commit perjury," said the book, quoting Charlemagne. "If two witnesses testify concerning some matter and disagree," said another section, based on the judgements of Louis the Pious,

---

[413] Capitularia regum Francorum, ed. Boretius, no. 41, §4, p. 117; no. 136, §2, p. 281.

[414] Judicial duels were, in fact, exceedingly rare in lawsuits of all kinds. See Elema, "Trial by Battle in France and England," 90-96, 134-9. Philip Augustus: Laurière, Ordonnances des roys de France de la troisième race, vol. 1, p. 35. For an example of bat-like clubs, see a late copy of the Coutumier de Normandie, Pierpont Morgan MS M.457, f. 85v. Baculi cornuti: Russell, "Accoutrements of Battle," 232-6; Elema, 249-50; Canel, "Le Combat judiciare en Normandie," 585-6.

[415] Fuero general de Navarra, ed. Ilarregui & Lapuerta, lib. V, tit. VII.1-3, p. 110.

"then let the count elect one person from one party and one from the other, and let those two fight with shields and clubs."

The customary law of Milan, recorded in 1215, said "Know that [a judicial duel] is always fought by means of champions with a shield on their head and a club (*fuste*), unless it is carried out otherwise by the consent of the parties." Around 1318, a Paduan writer thought club battles were used in even more kinds of cases: "The custom of the ancients was this: if two nobles or powerful men had engaged in a suit of homicide among themselves, each party was to find himself a hired champion, and on the ordained day these two champions ... [were] armed with shields, clubs and maces of wood (*clypeis, baculis et maschariis de ligno*)."

Even Holy Roman Emperor Frederick II's legal code for Sicily, the *Constitutions of Melfi*, said "Champions should have equal cudgels, neither spiny, nor spiked, nor having horns." In all of these Italian legal texts, swords were never mentioned as weapons for judicial combat. Sword duels would, however, arise later in the fourteenth century in the context of military tribunals, which had not developed as much formal evidential procedure.[416]

It is in this context that the most influential legal treatise of the Holy Roman Empire, the thirteenth-century *Sachsenspiegel*, is unusual in that it prescribes the use of swords for all participants in judicial duels (see fig. 3). Written sometime between 1225 and 1235, this text recorded the customary law of Saxony, a region then comprising much of what is now northern Germany. In criminal cases involving violence, an accuser could demand a duel from a man of equal standing or lesser status, and the judge was to supply a sword and shield to the accused if he needed it. According to the text, each combatant should have "a naked sword in hand, and one or two girded at his waist, as he prefers." He could switch from one weapon to the other during the combat. This book was highly influential on other German legal treatises of the thirteenth century. The *Deutschenspiegel*, and the *Schwabenspigel*, compiled respectively in Regensburg circa 1270 and in Augsburg around 1275, copied it nearly word for word in their own discussions of judicial combat. In northern Germany itself, how-

Figure 3: Dueling with swords and shields.
Legend: Sachsenspiegel, 1336. Rastede, Oldenburg Landesbibliothek, CIM I 410, fol. 34ͮ.

[416] Liber Papiensis: Leges Langobardorum, ed. Bluhme & Bonensi, p. 499, §65; p. 524, §3. Milan: Antiquitates Italicae medii aevi, ed. Muratori, vol. 3, col. 637. The dating of this last text comes from Hyde, Padua in the Age of Dante, p. 6. Frederick II: "...campiones habeant clavas equales, non spinosas, nec cum aguzonibus..." Historia diplomatica Friderici Secundi, ed. Bréholles & De Albertis de Luynes, vol. 4, part 1, 108-9. Military tribunals: Keen, The Laws of War in the Late Middle Ages, 19; Cavina, Il sangue dell'onore: Storia del duello, 47-9.

Figure 4: Dueling with swords and "longshields".
Legend: Gladiatoria, 1440s.
Krakow, Biblioteka Jagielloń-ska, Ms. Germ. quart. 2020, fol. 49ᵛ.

ever, trial by battle seems to have died out by the end of the thirteenth century, for there are no late medieval records of judicial duels.[417]

Subsequently, there are several examples of trials by combat with shields and swords from late medieval Swabia, which at the time comprised the upper watersheds of the Danube and Rhine Rivers, including Alsace and the Neckar River watershed. Although the *Schwabenspiegel* did not acquire its title until the seventeenth century, and was originally meant to describe German law in all imperial jurisdictions, its rules for duelling appear to have been particularly adapted to this region.

In Augsburg, the chronicler Ehrhard Wahraus noted that in 1409 he had seen two men fight a duel "behind shields" with swords and knives or messers in lists that had been constructed in the wine market. One of the men stabbed the other to death. In 1432 two men in Konstanz fought a duel with swords and tall shields. In Hans Talhoffer's manuscript from 1467, produced for Count Eberhardt of Württemberg, he begins the sequence depicting a duel with tall shields and swords with the words "Here I stand in the Swabian manner, as one does in [Schwäbisch] Hall" (see fig. 5). At the end of the Middle Ages, this last town gained a reputation for being one of the few jurisdictions in Europe which still allowed judicial duels, making it a magnet for bravos and a centre for early duels of honour.[418] Though records are few and centuries between, it seems that, unlike other parts of the Carolingian dominion, the former territory of the Alemannic law continued to consider swords the customary weapons of judicial combat throughout the High Middle Ages.

Not all German-speaking regions let duellists fight trials by battle with swords, however. Between the lands where Saxon legal customs dominated and the lands ruled by Swabian laws lay Franconia, a region with its own set of customs. Medieval

---

[417] "…en blot svert in der hant, unde en umme gegort oder tvei, dat stat an irme kore." Des Sachsenspiegels erster Theil, ed. Homeyer, book 1, tit. 63.4, p. 219. Der Schwabenspiegel oder schwäbisches Land- und Lehen-Rechtsbuch, ed. Lassberg, c. 79, §II.B, p. 38. Deutschenspiegel und Augsburger Sachsenspiegel, ed. Eckhardt & Hübner, 2nd ed., c. 88, §§7-8, p. 175. An exception to the disappearance of the judicial duel in Saxony seems to be the city law of Zwickau (on the Elbe watershed) from 1348. This may have to do with the town's proximity to Bohemia and Bohemian law. See the Zwickauer Rechtsbuch, ed. Ullrich, p. 26ff.
[418] Ehrhard Wahraus: Chroniken der schwäbischen Städte: Augsburg, vol. 1, p. 231. Konstanz: Chronicle of Christoph Schulthaiß, in Constantini M. Triarchus Triumphalis, etc., ed. Speth, 297-9 (I am grateful to Jens Kleinau for drawing my attention to this case). Cod. icon. 394a, op. cit., f. 65v. For the duelling ordinance of Schwäbisch Hall, recorded in 1537, see Württembergische Jahrbucher: 1843, vol. 2, ed. Memminger, 142ff.

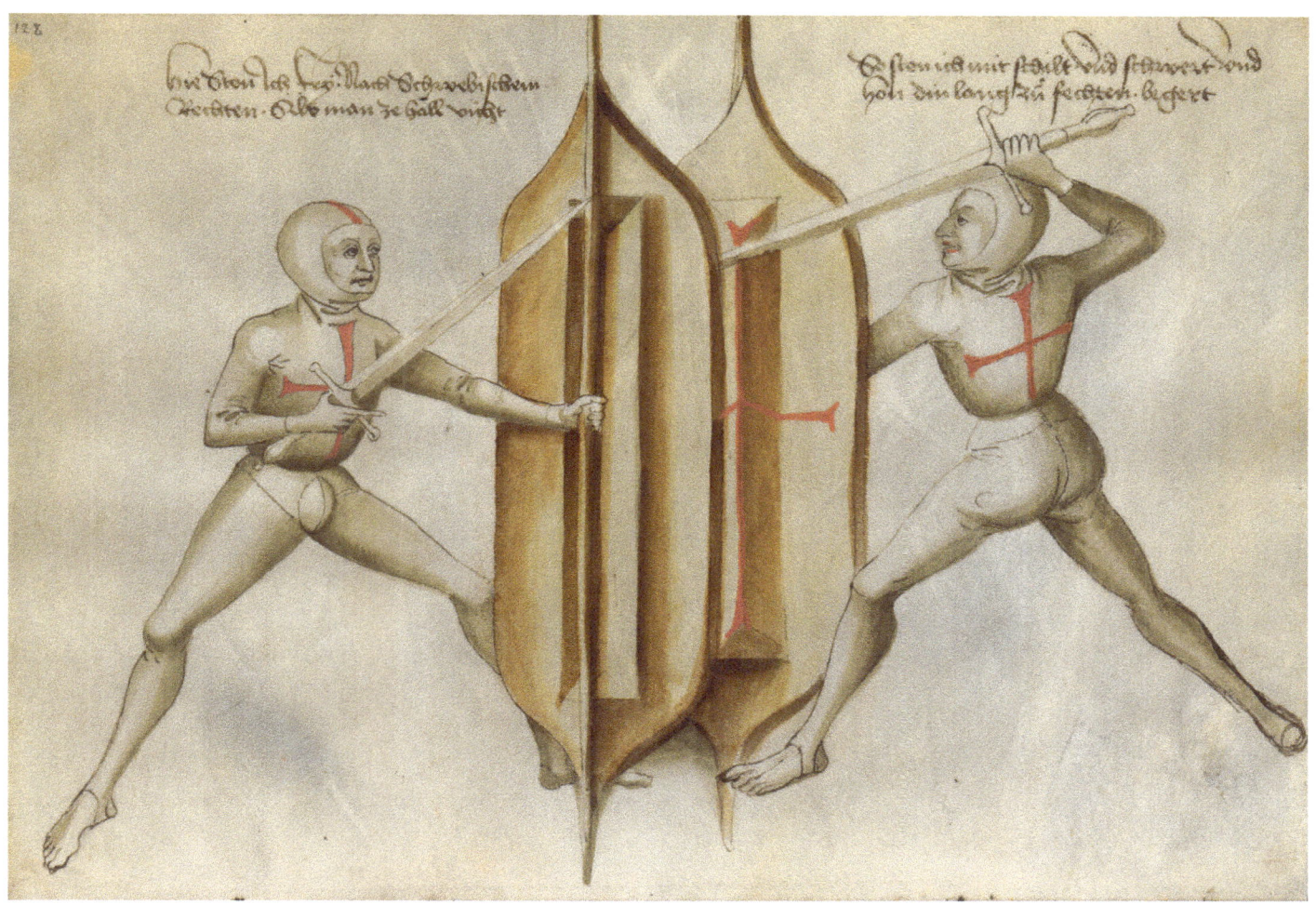

Franconia had different borders from its modern counterpart. It occupied the watersheds of the Main River and the Middle Rhine from roughly Worms to Bonn, including lands that had once belonged to the Ripuarian Franks. Many towns in this region began recording their oral legal traditions in writing in the thirteenth and fourteenth centuries, in documents known as *Weistümer*. Where these laws specify judicial duelling weapons, they always speak of clubs (*Kolben* or *Kolffen*). Such references can be found in the *Weistümer* of Alzey, from 1300; Bacharach, from after 1350; Gelnhausen, from 1360; and the general *Kampfgericht* of Franconia, recorded in the first half of the fifteenth century.[419]

Some of the most detailed duelling ordinances came from Franconian towns like Würzburg and Nuremberg, which lay at the edges of the Rhine watershed. These towns had frequent commerce with Swabian lands because they were the portage points to the Danube river system, and consequently they most likely needed their detailed legal texts to make clear the differences between their legal customs and those of neighbouring

Figure 5: The Swabian dueling form.

Legend: Hans Talhoffer, 1467. Munich, Bayerische Staatsbibliothek, Cod. icon. 394a, fol. 65ᵛ.

---

[419] Weisthümer, ed. Jacob Grimm, 7 vols. (Göttingen: Dietrich, 1840-1878), Alzey: vol. I, p. 799; Bacharach: II.213; Ordnung des Kampfgerichts von Franken: III.601-5. Gelnhausen: Hessisches Urkundenbuch ... vol 3: 1350-1375, p. 374. I am grateful to Jens Kleinau again for drawing my attention to the last ordinance.

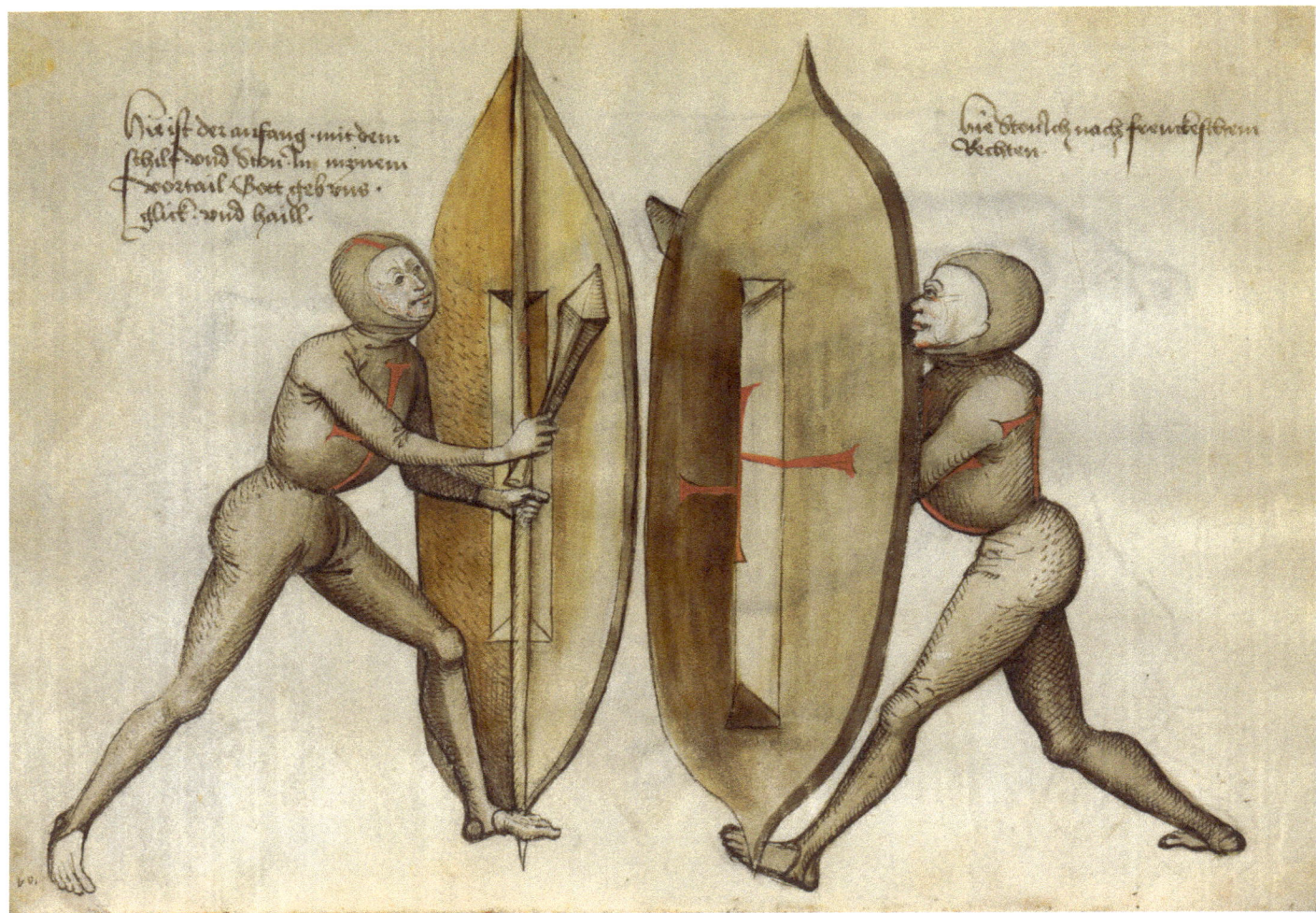

*Figure 6: The Franconian duel-
ing form.
Legend: Hans Talhoffer, 1467.
Munich, Bayerische Staatsbibliothek, Cod. icon. 394a,
fol. 53ᵛ.*

*Figure 7 (top right): Talhoffer's
club.
Legend: Hans Talhoffer, 1459.
Copenhagen, Det Kongelige
Bibliotek, Thott 290 2°, fol.
106ᵛ.*

Swabian lands.[420] In these Franconian towns, litigants all fought with clubs according to Franconian practice. At the beginning of Talhoffer's illustrations depicting men with clubs and tall shields in the Württemberg codex, he writes, "Here I stand according to Franconian law" (see fig. 6). The *Fechtbuch* of Peter Falkner from 1495 also begins its section on fighting with maces and shields with the words "Note that this is the Franconian combat"[421] (see fig. 18). The reforms to Frankish law enacted by Charlemagne more than six centuries previous were still visible in regional custom in Talhoffer's time.

It is hard to determine the origins of the particular type of club found in Talhoffer's manuscripts (see fig. 7). While illustrations of duelling clubs from other regions are not uncommon, they do not bear much resemblance to it. Talhoffer's

---

[420] The ordinance from Schwäbisch Hall comes similarly from the Swabian end of a portage. I am grateful to Anishinaabe tradition and more specifically to Dr. Niigaan Sinclair's discussion of the Selkirk Treaty Map and Cha Chay Pay Way Ti's Map at the following public lecture for impressing on me the importance of watersheds and portages in cultural relations: "Appropriation or Appreciation," OCAD University, 28 November 2018.

[421] Würzburg: "Ordnung des bitzing, ca. 1447," in Knapp, vol. I, part 2, p. 1281; Nuremberg: "Ordnung des Kampff-Gerichts des Burggraffgthumbs zu Nürnberg," in Teutsches Corpus Iuris, ed. Stephan, vol. 1, p. 709. Cod. icon. 394a, op. cit., f. 53v. "Merck daß ist der Frenckisch kampf..." in Peter Falkner, Kunste zu Ritterlicher Were, Vienna, Kunsthistorisches Museum, MS KK5012.

club has a heavy head like a mace and is four-sided in cross section. The only Franconian legal text to elaborate on the shape or dimensions of the clubs it prescribes is an ordinance from Würzburg, which is contemporary with Talhoffer. That document calls for the weapon to have three corners and a spike at the top, not unlike the illustrations in the master's treatises. Outside the corpus of *Fechtbücher*, the only analogous contemporary example known to this writer is the wooden club depicted in the tournament book of King René of Anjou, from 1460. It too has been planed down to faceted surfaces, but it is eight-sided in cross section and more regular in thickness. Unlike Talhoffer's clubs, it has also had its point cut off, and possesses a round guard (see fig. 8).[422]

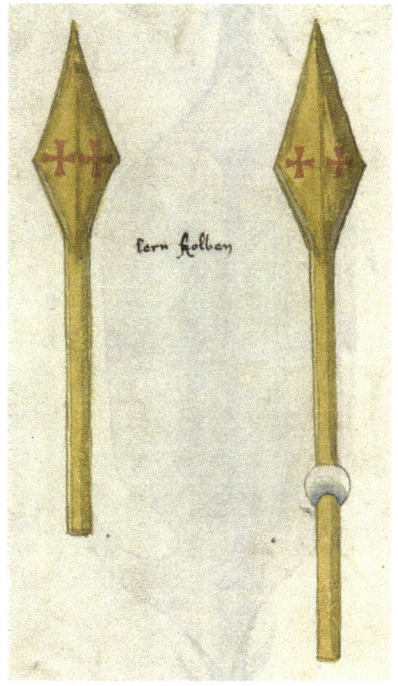

Figure 8 (below): Anjou's club. Legend: René d'Anjou, 1460. Paris, Bibliothèque nationale de France, MS Fr. 2695, f. 31v.

One may also note a less obvious resemblance between Talhoffer's clubs and the 'prismatic' iron and bronze mace heads found in medieval Transylvania and Hungary. These roughly contemporary devices were cylindrical in shape, growing thicker towards their tip, and octagonal in cross section, having eight flat faces without flanges. Their origins trace back to steppe cultures farther east, and they were also widely adopted by the Seljuk Turks in the Near East.

While Hungary and Anatolia may seem distant from Franconia, it must be remembered that they were only one portage away on a busy riverine trade route. The content of the treatise *Bellifortis*, which shares space with Talhoffer's work in one of his personal manuscripts, arrived in German lands by the same path on a similar timeline. Its author, Konrad Kyeser, was a physician who joined the disastrous crusade led by King Sigismund of Hungary against the Turks. After the defeat at Nicopolis in 1396 Kyeser's enmity with Sigismund led him to live in the mountains of Bohemia, where he fortuitously found illustrators to record the marvels of military technology he had seen abroad. An avenue for further research into the origin and spread of faceted clubs may be to investigate art and sculpture from Franconia specifically, but also the entire Danube trade route, in search of more analogues to these weapons.[423]

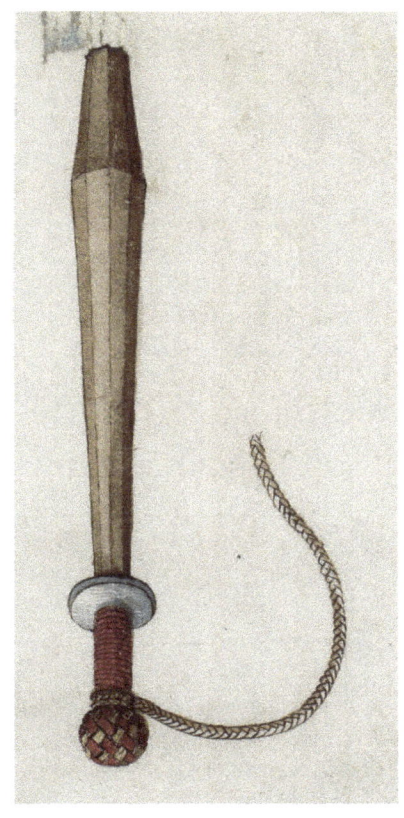

What can be said about the swords and clubs found in Talhoffer's duelling sequences is that we cannot assume that they were representative of judicial duels across Europe throughout the Middle Ages. Talhoffer depicted the weapons that could theoretically be used in his own period, in the regions with which he was familiar. Even within the fairly limited time and place that he operated, legal practices differed from one jurisdiction to the next, and the rules for trial by battle, though conservative, could absorb new developments.

---

[422] "...ieder kolb soll haben drei ecken u. fornen ein spitzen." "Ordnung des bitzing," in Knapp, vol. I.2, p. 1281. René d'Anjou, Traité de la forme et devis comme on peut faire les tournois, Paris, Bibliothèque nationale de France, MS fr. 2695, f. 31v.

[423] Fărcaș, op. cit.; Z. Boldog, "A prismatic mace-head from Dunaföldvár," 183. Kyeser: KB MS Thott, 290.2, 12r-48v; White, "Kyeser's 'Bellifortis'" 437.

Figure 9: Duel according to the Anglo-Norman form. Legend: Stained glass panel of Saint William, detail. York Minster, 1414.

Figure 10: Duel with swords and shields.
Legend: Sachsenspiegel, 1350-75. Wolfenbüttel, Herzog August Bibliothek, Cod. Guelf. 3.1 Aug 2°, fol. 26^r.

Shields

The shields for judicial duels described and depicted in medieval European sources show a great variety of forms, but most do not resemble the tall and unusually shaped specimens found in Talhoffer's manuscripts. Textual records are less helpful on this subject, as most of them simply prescribed that judicial duellists should use "a shield" (*scutum*, *schild*, or *escu*), without providing further detail. This ambiguity suggests that duelling shields were usually no different from other kinds of shields and needed no special description. Manuscript illustrations are more useful in providing specific information about their form.

From the thirteenth century onwards, French manuscripts usually depicted judicial combatants carrying triangular shields of the "heater" type, which were also common in military contexts.[424] In the Anglo-Norman tradition, duellists' shields were rectangular, ranging in size from early examples that covered fighters from shoulder to knee, to later targes that were only slightly larger than bucklers (see fig. 9).[425] These designs are unlikely to have been the direct ancestors of the south German type.

Notably, the shields described in thirteenth- and fourteenth-century German legal treatises also do not resemble the Talhoffer type. The *Sachsenspiegel* says quite clearly that a judicial duellist should have "a round shield in the off hand made of nothing but leather and wood, though the boss can be iron." The artist for the fourteenth-century copy of the manuscript now held in Wolfenbüttel interpreted this instruction as a large centre-held shield that would be slightly less than a metre in diameter in real life (see fig. 10). Another copy, from Dresden, illustrates the duellists holding shields a little larger than bucklers. The instruction to use round shields also carried over into the *Schwabenspiegel* and *Deutschenspiegel*, where it spread in many manuscript copies.[426]

---

[424] An exception is a marginal grotesque holding a horned club and a round red and white targe in the margin of a manuscript of the Decretals of Gregory IX from southern France, circa 1280. Tours, Bibliothèque municipale, MS 0568, f. 308. The image may be seen at http://initiale.irht.cnrs.fr/decor/9378, accessed 1 January 2019.
[425] An example of the large type can be found in a clerk's doodle on a Curia Regis roll from 1249, reproduced as the frontispiece of Select Pleas of the Crown, ed. Maitland. The smaller type is depicted in stained glass at York Minster and reproduced in French, York Minster: The Saint William Window, plate 10. See also Russell, "Accoutrements," 436-7.
[426] Sachsenspiegel: "Enen senewolden schilt in der anderen hant, dar nicht denne holt unde leder an ne si, ane die bokelen, die mut wol isern sin." ed. Homeyer, p. 219. Wolfenbüttel, Herzog August Bibliothek, Cod.

Meanwhile, most of the Franconian *Weistümer* that discuss trial by battle simply call for duellists to use "*ein schild*," which likely meant they assumed this piece of equipment was no different in design from a common military shield. The duelling ordinance of Nuremberg from 1410 instructs that the shield should be made from nothing but wood, lime and sinew, and covered with a white linen cloth marked with a red cross. It does not, however, say anything about the shield's size or shape. Nevertheless, the custom of Bacharach, written before 1350, calls for duellists to be prepared "*myt syme roiden schilde*," another reference to a round shield.[427] Consequently, not only are there no examples of tall shields before the fifteenth century, there is even evidence that duellists in German lands were using shields of an entirely different shape.

When did the "longshield" (as one of the *Gladiatoria* manuscripts dubs it, see fig. 4) first appear and where did it come from? An intriguing clue comes from Bohemia. A bilingual Latin-Czech manuscript of Bohemian law, the *Ordo judicii terrae*, or *Řád práva zemského,* contains glosses written sometime after the death of Emperor Charles IV in 1378. One gloss, on a passage about a judicial duel between a burgher and a peasant, says "According to the law of old, they used to duel with clubs and small shields carried by weapons-bearers, but now they may fight with swords and great shields."[428] This passage would place the appearance of the longshield in Bohemia in the third quarter of the fourteenth century or later.

The Bohemian law accords with scattered scraps of evidence from south German lands. The first unambiguous evidence for tall shields I can find is the account of the duel between Hans Roth and Hans Riem at Konstanz in 1432 over an accusation of malicious sorcery. In this case, both men had a shield that reached their head (*Kopff reichende Schild*). At roughly the same time (circa 1430) an anonymous treatise on arts and wonders included an image of two men fighting with spiked clubs and rectangular shields sporting cone-shaped central bosses and sets of three spikes protruding from their upper and lower rims (see fig. 11).

Figure 11: Dueling with swords and shields.
Legend: Kunst und Wunderbuch, 1430. Rome, Biblioteca Apostolica Vaticana, Cod. Pal. Lat. 1888, fol. 91ᵛ.

Guelph. 3.1 Aug. 2, f. 26r; Sächsischen Landes- Staats- und Universitätsbibliothek Dresden, Mscr. Dresd. M.32, f. 20r. Schwabenspiegel, ed. Lassberg, op. cit. p. 38; Deutschenspiegel, op. cit., pp. 175-6.
[427] "Ordnung des Kampff-Gerichts des Burggraffgthumbs zu Nürnberg," op. cit., § 11, p. 709. Bacharach: Weisthümer, ed. Grimm, II.213.
[428] "Řád práva zemského - Ordo iudicii terrae," in Codex juris Bohemici, ed. Jireek, vol. 2, part 2, p. 198. "Et juxta antiqua jura cum baculis et clypeis parvis, quos hastiferi deferent, duellabant; nunc autem, duellare debent cum gladiis et magnis clypeis..." Ibid., p. 219, § 29.

The only German legal document to mention a shield of the Talhoffer type is the *Ordnung des Bitzing* of 1447 from Würzburg. That ordinance calls for the use of "a shield that has three spikes on each side and is as long as the man [who carries it]." (One version of the Nuremberg duelling ordinance also allows for spikes on the top and bottom of the shield.)[429] It is therefore likely that in the first half of the fifteenth century, tall shields were a relative novelty that had appeared in south German lands within living memory, perhaps even within Talhoffer's own lifetime.

There is, however, a much older precedent for the Talhoffer shield. Nearly two centuries before the duel at Konstanz, tall shields and shields with spikes made an appearance in the *Livre des assises de la cour de Barons,* which was part of the collection of legal treatises known as the *Assizes of Jerusalem.* John of Ibelin, Count of Jaffa, wrote this treatise between 1264 and 1266 to record the customs of the high court of the crusader kingdom of Jerusalem as they had existed before Saladin reconquered that city in 1187. In the section about cases of homicide where both the accuser and the accused are knights, this treatise calls for a judicial duel on foot in which each man is equipped with "a shield which one calls a *harace,* which is larger than him by half a foot or a full palm, and has two piercings together, in such a place that one can see one's adversary through these holes..." In other kinds of judicial battles between knights, "the shield should have two spikes, one in the middle of the shield and one below on the foot, and they should be of any size and length [the knight] wants, up to a foot long but not more, and the shield can have as many sharp iron spikes or blades as he wants."[430] In 1369, this treatise was also adopted as the official reference book for the high court of Cyprus, although no judicial duels seem to have been held there after that date.

Shields like these are not found in any other European legal treatises of the thirteenth or fourteenth centuries, until the brief reference in the gloss of the Bohemian *Ordo iudicii terrae.* They seem to have developed in the Middle East. Arabic military treatises of the time do not contain direct analogues, but they do mention some shields from which the *harace* may have derived. The late twelfth-century *Tabṣira,* written for Saladin by the engineer Murḍa bin Alī al-Tarsūsī, describes two kinds of shields used by the Franks. The *tariqa* was a long

Figure 12: Shield and club for dueling.
Legend: Assises de Jerusalem. After Oxford, Bodleian Library, MS Selden Supra 69, fol. 104ʳ.

[429] Constance: Chronicle of Christoph Schulthaiß, ed. Speth, op. cit., p. 298. There was also the duel "behind shields" mentioned by Wahraus, op. cit., 231. Spiked shields: Vatican, Biblioteca Apostolica MS Pal. Lat 1888, f. 89v. Würzburg: "Item: ein schilt, der uf ieder seiten hat drei spitzen u. als lang der man ist." Ordnung des bitzing, in Knapp, p. 1281 Nuremberg: in Teutsches Corpus Iuris, ed. Stephan, vol. 1, p. 714, § 50.

[430] "...une targe, que l'on apele harace, qui soit plus grant de lui demi pié ou plaine paume, et laquel ait .ii. pertuis de comunal grant en tel endroit qu'il puisse veir son aversaire par ceaus pertuis." John of Ibelin, Le Livre des Assizes, ed. Edbury, tit. 89, p. 236. "...et en l'escu doit avoir .ii. broches, l'une en mi l'escu et l'autre au pié desout, et doivent estre de tel groisse con il vodront et de tel longor jusques a .i. pié mais neent plus." Ibid., tit. 90, p. 240.

shield that was round on top but gradually narrowed to a point, probably representing the "kite" shield popular in Europe at the time. The *januwiya*, on the other hand, was similar to the *tariqa*, but flat on the bottom, so that foot soldiers could rest it on the ground and create a shield wall resistant to archers. It seems to have been an early form of pavise, and indeed the word most likely derives from *Genoa*, since it does not have a clear etymological root in Arabic.[431] Scaled slightly larger and pierced with holes, this latter shield may have become the *harace*.

Despite the time and geographical distance separating the *Assises de Jerusalem* from the German fight book corpus, there are notable similarities between these books. One manuscript of the *Assises* from the first half of the fourteenth century has a line drawing in the margin of a knight carrying a spiked shield as described in the text. It is depicted as a heater type with a spike protruding from its lower rim, and concentric circles representing the second

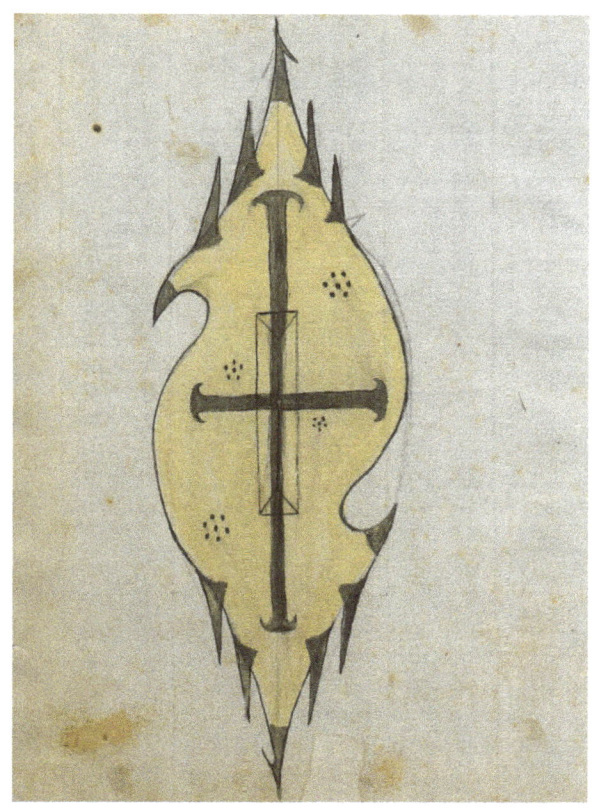

spike at its centre boss (see fig. 12) This shield appears to be decorated with three clusters of five dots in the shape of flowers. A long shield with many spikes depicted in the earliest of the Talhoffer manuscripts, dating from circa 1448, is also decorated with a similar motif: four clusters of five or six dots are arranged around central dots in flower-like patterns (see fig. 13). Another German treatise, the so-called *Cluny Fechtbuch* created in the last quarter of the fifteenth century, shows two figures fighting with tall shields bored through with a matrix of piercings (see fig. 14), not unlike the ones on the *harace* described in the *Assises*.[432]

Many more duelling shields in the corpus of German *Fechtbücher* have spikes or blades protruding from their rims. Although none of the five surviving manuscripts of the *Assises de Jerusalem* are known to have sojourned in German lands during the Middle Ages, it may be that a copy of the text—or perhaps the oral tradition that it represented—found its way to the Rhine-Danube network towards the beginning of the fifteenth century.

The transmission of specialized duelling shields from the Mediterranean world to Bohemian and south German lands mirrors the better-documented history of pavises generally. In Italy, the long kite shield was not supplanted by the triangular heater shape in the thirteenth century

Figure 14: Dueling shield. Legend: Talhoffer, 1448. Gotha, Forschungsbibliothek Erfurt/ Gotha, MS Chart.A.558, fol. 24ᵛ.

Figure 13: Dueling with clubs and shields. Legend: Paris, Musée national du Moyen Âge, Cl. 23842, fol. 180ʳ.

[431] Claude Cahen, "Un Traité d'armurerie composé pour Saladin," 137; 156, n. 3.
[432] Shield pictured in the Assises: Oxford, Bodleian Library, MS Selden Supra 69, f. 104r. Shield in Talhoffer: Gotha MS Chart. A.558, f. 24v. Cluny Fechtbuch: Paris, Musée national du Moyen Âge, MS Cl. 23842, f. 180r.

Figure 15: Seal of Henri Chaillau, featuring a spiked shield and horned club. Legend: 1300s. After Paris, Archives nationales, Collection de sceaux, no. 5860.

Figure 16: Apes with notched kite shields. Legend: Book of the Apocalypse, 1313. Paris, Bibliothèque nationale de France, MS fr. 13096, f. 32ʳ.

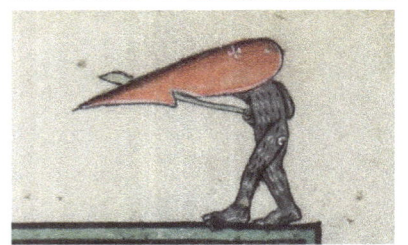

as occurred farther north. A larger, more rectangular version of the kite shield, behind which archers could shelter, was first mentioned in the thirteenth century, and popularized by companies of Italian mercenary archers who brought it to France during the Hundred Years' War. German writers of the time also knew of a large rectangular shield that rested on the ground, calling it a *Setzschild* or *Setztartsche*. The Hussite wars of 1419 to approximately 1434 sparked the development of a distinctively Bohemian form of the pavise with a vertical rib down the length of its centre line, and occasionally a wooden hook or beak in the centre of its top edge.[433] The tall German duelling shield, with a vertical rib-like boss and often hooks at its head and foot, was likely an offshoot of the same influences, as soldiers and materiel moved through the upper Danube on their way to and from the wars.

Outside of southern German lands and the *Assises de Jerusalem*, specialized duelling shields occurred only as isolated outliers. In the Archives Nationales de France, a seal dated broadly to the fourteenth century depicts a duellist's club with horns superimposed over a heater-type shield bearing a spike in the centre of its rounded lower rim (see fig. 15). From behind the centre of the shield's upper edge, a shaft and spearhead protrude. Two other lines emanating from the shield's upper edge may represent either decorative curlicues or hooks that were part of the shield. The legend around the rim reads "Seal of Henri Chaillau, the fencer of Châlons."[434] Although horned clubs are found in some sources from the French-speaking realm, this appears to be the only French reference to a spiked duelling shield.

In another case, a marginal drawing in a Book of the Apocalypse shows apes facing off with clubs and what may be early Talhoffer-type shields (see fig. 16). The shields are kite-shaped and as tall as their wielders, an unusual form to find in a French manuscript dated to 1313. Each shield has a large notch in its rim, resembling a lance rest but positioned on the wielder's lower left side rather than the upper right corner.[435] The manuscript was created for Isabella of France, Queen of England, most likely to commemorate her visit to Paris that year along with her husband, Edward II, on which occasion her

---

[433] Denkstein, "Pavises of the Bohemian Type, II," 167-9, 172-9, 184-5. DeVries, "The Introduction and Use of the Pavise in the Hundred Years War," 98.

[434] "S. HENRI CHAILLAV L'ESCREMISSEEVR DE CHAALONS," Douet d'Arcq, Collection de sceaux, part 1, vol. 2, p. 396, no. 5860. Douet d'Arcq thought the shield was a steaming cauldron and the club was a fire rake, but in light of the seal's legend, this interpretation is unlikely.

[435] Paris, BnF, MS fr. 13096, f. 32r.

father reaffirmed his commitment to go on a crusade.[436] To my knowledge, no other illustration of French or English shields depicts them as having that notched shape.

As was the case for the prismatic club and the Near East, there are examples in one of Talhoffer's personal manuscripts of knowledge transmission from the Middle East and the greater Mediterranean world. In the pages following Talhoffer's own treatise, the Thott manuscript contains a treatise on the anatomy of internal organs by a writer who is referred to as the Jew Ebreesch (i.e. Hebraic), who cites a "Maister All-monser" and his book "panthagin". As Diether Bachmann has noted, these are references to Manṣūr ibn Ilyās (1380-1422), a Persian physician who wrote a treatise on anatomy, and the separate medical treatise known as the *Liber Pantegni*, written by Constantine the African, an eleventh-century Tunisian physician who migrated to Italy and became a monk at Monte Cassino. In addition, there follows an explanation of Hindu-Arabic numerals, a reckoning system that, while known to a few European mathematicians since Fibonacci, was only just becoming popularized in Germany.[437] Clearly, Talhoffer was collecting specialized technical knowledge from well outside his local sphere.

Consequently, in south German lands, specialized duelling shields appear to have been a short-lived phenomenon that existed mainly in the first half of the fifteenth century. Rather than having an ancient or particularly Germanic origin, they followed flows of migration from the Middle East to their new home. After their appearance in the Würzburg ordinance of 1447, they were only ever mentioned in the *Fechtbücher*. However, among the German fighting masters their popularity continued for several generations more, perhaps as a late medieval form of re-enactment.

Clothing

In thirteenth-century legal treatises, clothing for judicial duelling was not much different in style from other contemporary apparel. The illustrations in the Wolfenbüttel and Dresden manuscripts of the *Sachsenspiegel* show duellists fighting in hose and loose knee-length tunics. The text says "They should put on as much leather and linen apparel as they want. Head and feet are bare in front, and on their hands they should have nothing but thin gloves. ... A jerkin without sleeves goes over the gear." These instructions were copied virtually word-for-word in the *Schwabenspiegel* and the *Deutschenspiegel*. A thirteenth-

---

[436] Paris, BnF, MS fr. 13096, f. 32r. Bergot, "L'Apocalypse d'Isabelle de France (1313)," 70. Suzanne Lewis, "The Apocalypse of Isabella of France: Paris, Bibl. Nat. MS Fr. 13096," 226-7.

[437] KB MS Thott 290.2, ff. 141v, 149v, 150v; "Jud Ebreesch," Wiktenauer, note 6, https://wiktenauer.com/wiki/Jud_Ebreesch#cite_note-6, accessed 1 January 2019; Arabic numerals: Hill, pp. 42-3, tables XXIII-XLIII, pp. 62-95.

century recension of the *Sachsenspiegel* originating in Augsburg substitutes wool for linen and adds that each of the combatants should have not only as much apparel as he wants, but also as much as his opponent wears, but those treatises keep the rest of the passage intact.[438]

Furthermore, it was not clear from the treatises of this period that knights should fight judicial battles any differently than burghers or peasants, for there is no mention of armour. The only thirteenth-century German treatise to allow steel apparel was the city law of Colmar in Alsace, which permitted both combatants a hauberk (*hassberch*).[439] Unlike in France, where knights were permitted to fight judicial duels in full armour on horse-back, thirteenth-century German customals reflected a society where knights were not necessarily much wealthier or more influential than burghers.

Figure 17: Clothing for dueling.

Legend: Hans Talhoffer, 1459. Copenhagen, Det Kongelige Bibliotek, Thott 290 2°, fol. 107ʳ.

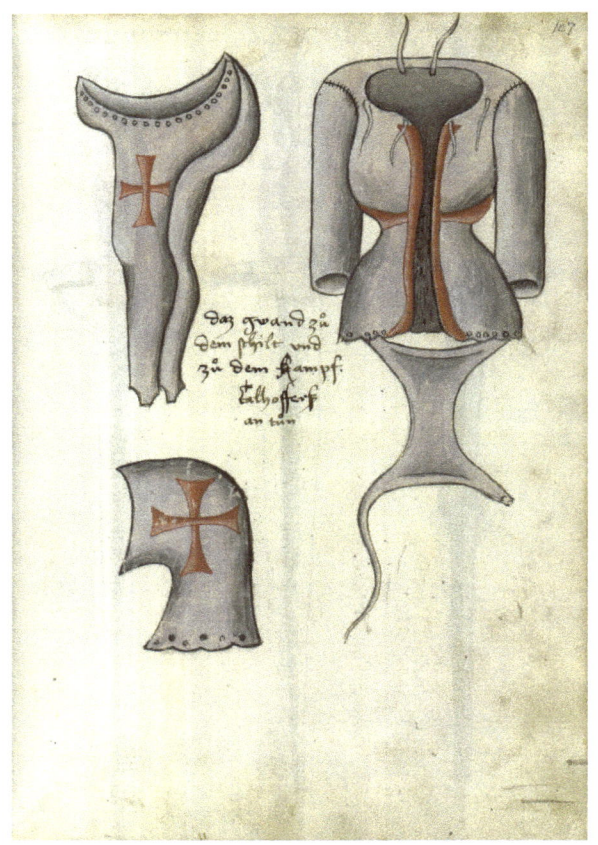

Rules for clothing appear to have been much the same as Saxony in Franconia. A *Weistum* from Bacharach dating from before 1350 says that a combatant should appear with "his jerkin taken in, ... with his white felt, with his hat bound on, with all those things that are proper for combat.[440] The jerkin may be an example of the tightly fitted clothing found later in the fighting treatises, or it may simply have been hiked up and belted in. Fitted clothing was also making an appearance in images of duellists in France and England at this time.[441] References to felt also turn up in French and English documents about trials by battle. It may have been used to provide some degree of padding.[442]

The slightly later duelling rules from Gelnhausen are the first to introduce two elements that would later appear in fight books from both Franconia and Swabia. The duellists' clothing was to be grey, and it was to have leather crosses attached in front and behind. This development appeared not long after the Black Death. Together with the rules about fighting barefoot, which had existed

---

[438] "Leder unde linen ding muten se an dun, alse vele alse se willet. Hovet unde vüte sint in vore blot, unde an den henden ne solen se nicht wen dunne hantzeken hebben; ... ene senewolden schilt in der anderen hant, dar nicht denne holt unde leder an ne si, ane die bokelen, die mut wol isern sin, enen rik sunder ermelen boven der gare..." Sachsenspiegel, ed. Homeyer, book 1, tit. 63.4, p. 219. Schwabenspiegel, ed. Lassberg, §79.II.B, p. 38; Deutschenspiegel und Augsburger Sachsenspiegel, Landrecht, c. 88, § 8, p. 175.

[439] "Das Recht der Stadt Colmar," in Deutsche Stadtrechte des Mittelalters, ed. Gaupp, vol. 1, § 39, p. 121.

[440] "...syme einfaren rocke, ... mit syme wissen viltze, iyt syme uffgebunden huote, myt alle deme, daz man zum kampfe begeert..." Grimm, Weisthümer, vol. 2, p. 213.

[441] See, for instance, the late thirteenth-century French duellists in Berlin Staatsbibliothek MS Hamilton 193, f. 194r, or the early fourteenth-century English ones in the Smithfield Decretals, London, BL, MS Royal 10 E IV, f. 96v.

[442] Russell, "Accoutrements," 437-8. Elema, 252-3.

since the thirteenth century, these elements gave an air of penitence to the combatants, who were in some danger of causing the death of their opponent or meeting their Maker themselves. Grey clothing for duellists is also prescribed in the Franconian ordinance on duelling law from the first half of the fifteenth century. A Swabian account of a duel at Hall in 1405 describes the combatants as being dressed in ash-grey (*leucophaeo vestitu*), recalling the biblical expression "sackcloth and ashes". The Nuremberg duelling ordinance of circa 1410 also calls for crosses on clothing and shields, an echo of the ones worn by those who had vowed to go on crusade. In this context, some of the odder clothing choices in the fighting treatises, such as the combatants with "balaclava" headgear in Falkner and the Cluny Fechtbuch, make more sense. Similar hoods appear in a French miniature from the early fifteenth century depicting a procession of flagellants.[443] They represent the shame and penitence of their wearers.

Of course, most German *Fechtbücher* also included a section on duels in armour. These duels arose out of a slightly different legal tradition. Even as each geographical jurisdiction had customary laws that were beginning to be codified, Western Christiandom as a whole was starting to develop a body of law about how war was to be conducted and how fighting men were to resolve disputes among themselves. Military commanders had traditionally possessed some powers to arbitrate disputes among their troops. With the increasing use of foreign mercenary companies in the fourteenth century, international consensus began to develop the *ius armorum*, or customary law of arms. It came to be agreed that, after an army had disbanded, cases under this law could be heard in any court of a sovereign prince, regardless of the court's location or relation to the original crime. The law of arms regarded judicial duels more favourably than other customs did, and also permitted litigants to fight their battles in armour and on horseback, since all participants were already men at arms, and often aristocrats, who possessed such equipment already.[444]

Medieval chroniclers tended to discuss these armoured duels alongside other single combats of a less judicial nature. The

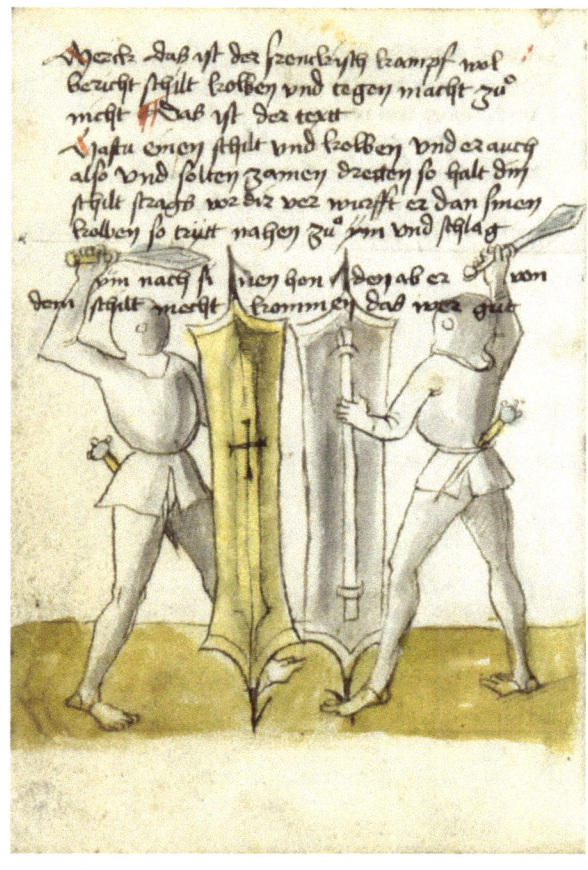

Figure 18: Dueling with clubs and shields.
Legend: Peter Falkner, 1495. Vienna, Kunsthistorisches Museum, MS KK5012, fol. 65ᵛ.

---

[443] Gelnhausen: Hessisches Urkundenbuch, op. cit., 374. Franconia: Ordnung des Kampfgerichts von Franken, op. cit., p. 601. Duel at Hall: Crusius, Annales Suevici, book 3, p. 330. Nuremberg: "Ordnung des Kampff-gerichts," op. cit., § 11, p. 708-9. Falkner, 65v-67r. Cluny Fechtbuch: MS Cl. 23842, f. 181r. Flagellants: Belles heures de Jean, Duc de Berry, New York, Metropolitan Museum of Art, Cloisters Collection MS 54.1.1, f. 74v.
[444] Keen, Laws of War, 19-20, 41. Elema, 76-81.

Figure 19: Armored dueling. Legend: Hans Talhoffer, 1459. Copenhagen, Det Kongelige Bibliotek, Thott 290 2°, fol. 132ʳ.

Hundred Years War in France saw the rise of the deed of arms, a formal, arbitrated and regulated combat that blurred the distinctions between tournaments, warfare, and judicial duels. This phenomenon also appeared in the Holy Roman Empire as early as 1336, when Emperor Ludwig IV arbitrated a duel between his chamberlain Hector von Trautmannsdorf and the knight Seyfried der Frauenberger over the question of whether the latter "slanders his [Hector's] honour behind his back and claims himself to be better and to come of a better and nobler lineage than him." The Kaiser originally tried to settle the dispute by having both men produce documetation of their lineage, but when Hector was not satisfied with Seyfried's document, the Kaiser agreed to grant them a day to fight. The men pledged their shields and helms, implying that the duel was to be held in armour, and agreed to surrender their arms to their opponent if they were defeated, a practice up to then associated with tournaments rather than judicial combat.[445]

In 1386 the French prior Honorat Bovet produced the *Arbre des batailles*, a book which included a codification of the rules for armoured duels. Christine de Pizan expanded on his work in 1413, making special mention of the German empire. The

[445] On deeds of arms in general: Muhlberger, Deeds of Arms, 6-8. On Trautmannsdorf v. Frauenberger: Müller, Reichs Tags Theatrum, pp. 103-4.

nineteenth-century scholar Alwin Schultz collected ten cases of late medieval German duels dating between 1347 and 1478, though at least one of them was fought in grey clothing without armour. These late medieval duels and the references to armour in them would benefit from a more in-depth study and a comparison to the fight books, although it is beyond the scope of this article.[446]

Thus, duellists' clothing or armour was a marker for two separate but related legal traditions. Those who fought judicial combats barefoot and without armour were following an older set of customary procedures rooted in local land law, while those who fought in armour were following a younger, military-based custom that had developed in a wider West European context. These two sets of customary laws coexisted and sometimes even overlapped in their jurisdictions, a contradiction that bothered medieval jurists far less than their modern counterparts.

Duels between a man and a woman

One more unusual kind of judicial duel also occurs in the manuscripts of Talhoffer and his successors: the duel between a man and a woman. In this scenario, the man fights with a club while standing in a waist-deep pit, while the woman moves freely around him, armed with a stone tied in a length of cloth. The duel occurs as a sequence of illustrations in Talhoffer and Paulus Kal, while the anonymous book of arts and wonders, Hutter and the Cluny Fechtbuch include it as a single image.[447]

The website *Wiktenauer* refers facetiously to duels of this kind as "Marriage Counseling," but when they are described in medieval legal treatises, they are always cases of rape. The medieval term for this crime was *Notnumpht*, a word which could also encompass kidnapping and associated assault. The earliest legal treatises that prescribe judicial duelling as a remedy for it were written in a relatively confined region around Munich. The city law of Augsburg from 1276 says that if there were no witnesses who saw or heard the rape, the woman could bring an accusation against the rapist in court, but he could clear himself by swearing single-handed, that is, by swearing a simple oath to his innocence without having any associates to join him as oath-helpers. If, however, the woman was unwilling to accept his oath, then she was obliged to fight him in person. The man was to stand in a

Figure 20: Duel between man and woman.
Legend: Paulus Kal, 1470. Munich, Bayerische Staatsbibliothek, Cgm 1507, fol. 49ᵛ.

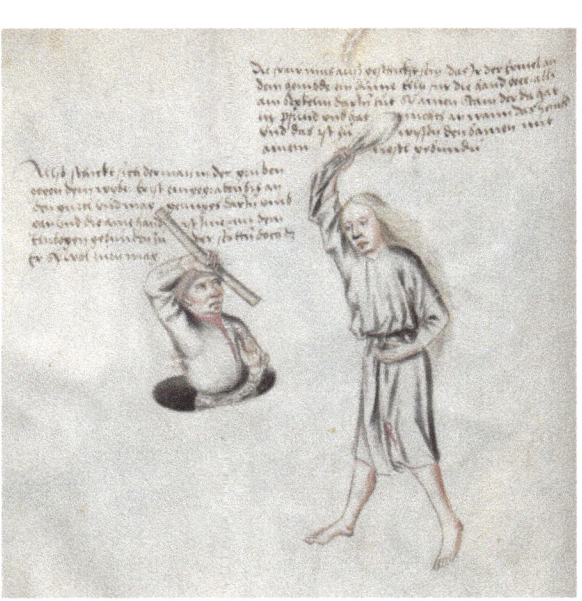

---

[446] Bovet, ed. Nys, 222-46. Christine de Pizan, ed. Willard, 199-200. Schultz, *Deutsches Leben*, 369-70.
[447] Talhoffer MS Thott 290.2, ff 80r-84r. Kal, MS Cgm 1507, ff. 49v-51v. Arts and Wonders: MS Pal. lat. 1888, f. 94r. Cluny Fechtbuch, MS Cl. 23842, f. 194v. Hutter, MS Cg. 3711, f. 58a.r.

hole up to his navel, armed with an oaken cudgel, while the woman was armed with a small garment (*roeclin*) in which there was a fist-sized stone. Whichever of the two was defeated was to be buried alive.[448]

Ruprecht von Freising's *Freisinger Rechtsbuch* of 1328 calls for much the same scenario, except it specifies that the man's left hand should be bound behind his back and the woman's stone should weigh one pound. In this case, the man who lost was to be beheaded, while a woman who lost was to have her hand chopped off. The *Rechtsbuch* of Kaiser Ludwig IV the Bavarian from 1346 also mentions a duel between a woman and a man in cases of rape, but it provides no details about how the combat is to be carried out.[449] The conditions laid out in these texts seem to have been designed to discourage accusations of rape from being brought before the courts. Needless to say, there does not seem to have been any case law from Bavaria in which a duel of this sort was ever carried out in real life.

There was, however, one duel between a man and a woman held some distance away, in Bern, in 1288 (see fig. 22).

The laconic chroniclers of the time gave the incident no more than one sentence each. The late thirteenth-century annals of Colmar reported for that year "In the city of Bern, a

Figure 21: Duel between man and woman.
Legend: Hans Talhoffer, 1459. Copenhagen, Det Kongelige Bibliotek, Thott 290 2°, fol. 80ᵛ.

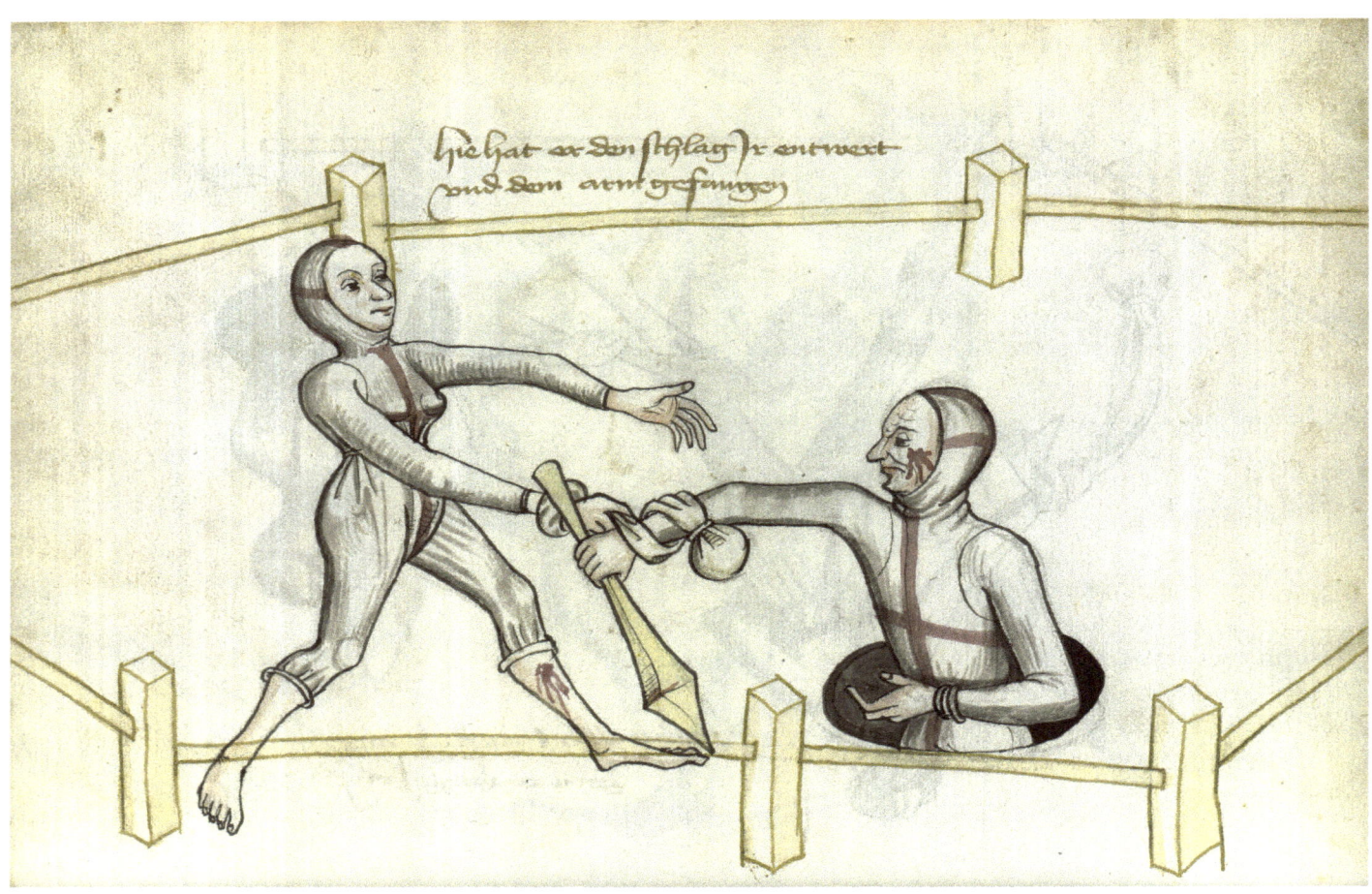

---

[448] Stadtbuch von Augsburg, ed. Meyer, art. 31, § 1, pp. 89-90.
[449] Ruprecht von Freising, ed. Claußen, § 137. Rechtsbuch Kaiser Ludwigs, ed. Volkert, art. 58, p. 291.

woman is known to have defeated a man in a duel." The *Chronica de Berno* from circa 1325 said "In the year of the Lord 1288 there was a duel in Bern between a man and a woman on the octave of the Innocents, but the woman prevailed." Conrad Justinger's *Berner Chronik* had little more to add in 1420: "Men say that in 1288, on the octave of the Innocents, a battle occurred in the market where the churchyard wall now stands; a man and a woman fought each other and the woman won." From these sources, the story was copied into Diebold Schilling's *Spiezer Chronik* and the sixteenth-century chronicle of Johann Stumpf. This is the sole example in European history of a trial by battle between a man and a woman that the litigants pursued all the way to combat and fought in person. However, none of the chronicles provides the combatants'

names or describes the nature of the suit or the weapons in any way.[450]

Figure 22: Duel between man and woman.
Legend: Diebold Schilling, Spiezer Chronik, 1484/5.
Bern, Burgerbibliothek, Mss. h.h.l.16, p. 112.

The duel in Bern may have spread the fame of the procedure. Around 1300, the poet Heinrich von Neustadt from Vienna included an episode of a woman duelling a man in his epic poem *Apollonius von Tyrland*. This poem was one of several versions of the Apollonius story circulating Europe at the time, but none of the others contained this incident. In Heinrich's version, the woman challenges the man because he has slandered and attempted to violate her sister. Her weapon is a stone weighing three pounds sewn into a cloth two ells long. The man, as per the Bavarian tradition, stands waist-deep in a narrow pit with one hand tied behind his back.[451]

By Talhoffer's time, the procedure seems to have become a theoretical possibility in some Franconian courts. An ordinance from Würzburg dating to circa 1447 says that in a duel where a woman accused a man, the woman should have three maces, each one made from a shaft of hazel one ell long, with a one pound head of stone attached to it with leather thongs. As in other cases, the man was to stand in a hole. He was armed with three clubs, each one an ell long and as thick as two man's thumbs. Every time he struck at the woman and his hand

---

[450] The text of the entries in the Annales Colmarienses, Chronica de Berno and Berner Chronik is recorded in Neumann, Der gerichtliche Zweikampf, p. 206, n. 1118. Schilling, Spiezer Chronik, p. 112. Johannes Stumpf, Gemeiner loblicher Eydgnoschafft, book 8, cap. 6, f. 250v. A surviving Bernese legal customal from seventy years before the incident does not mention duels between a man and a woman, but does say that trial by battle can be used in cases of secretive murder and violent nocturnal home invasion. Zeerleder, Urkunden, vol. 1, p. 185.
[451] R.W. Pettengill, "The Source of an Episode in Heinrich von Neustadt's Apollonius," 45-6.

touched the ground, he was to lose a club. If he lost three clubs, he would lose the battle. Likewise, if the woman struck at the man and missed, she would lose a club, and if she lost the battle she could be buried alive.[452] Although German chronicles were more plentiful and detailed in the fifteenth century than in the thirteenth, there are, unsurprisingly, no records of anyone holding such a battle in earnest.

Nevertheless, case records exist in which battles between a man and a woman were threatened. In 1405, Els, daughter of Otto Spara of Schwarzach, sent Albrecht Lotter of Wiernsreut a summons to answer for an unspecified offence. "If he confirms it then she will be satisfied, but if he denies it, she wishes to make it true with her stone on his head, according to the law of battle." The case was delayed three times, after which a further ten months went by before the clerk added "Judged in Fürth after [the feast of Saint] Vitus in the year 1406." Another case in 1435 ends with the words "it has been judged with law (*mit recht verurteilt*) on the Monday before Maundy Thursday."[453] This last remark most likely means that the case was settled according to a process recorded in academic written law, rather than a customary law procedure like battle. A judicial duel in either case would surely have caused enough sensation to have been noted by one of the many chronicles that recorded happenings in the Nuremburg region.

These precedents raise important questions about the combats portrayed in the *Fechtbücher*. It is unlikely that either Hans Talhoffer or Paulus Kal ever witnessed a judicial duel between a man and a woman or talked to anyone who had seen one. How then did they learn the plays they illustrated? Did they, like the historical martial artists of today, reconstruct the fight from scraps of historical knowledge and the application of physical principles? How did they go about this work? Furthermore, if today's martial artists attempt to recreate the plays of Talhoffer and Kal, are they in fact reconstructing a re-enactment?

Conclusion

Until now, most historians have attempted to study German judicial duels from a national perspective. This focus is at once too broad and too narrow. German trials by bat-tle were rooted in regional customary law that predated the Holy Roman Empire. Even in the fifteenth century, the autumn of this procedure's history, regional custom determined how it was carried out, particularly the rules for the weapons and

---

[452] "Ordnung des bitzing," in Knapp, vol. I.2, p. 1282.
[453] "...iehe er ir das, das sey ir leib, laugne er ir aber sein, so will sie es mit irem stein auf sein haubt war machen nach Kampfsrecht." In Jung, Miscellaneorum, vol. 1 p. 187. Knapp, vol. 2, p. 418, n. 5.

equipment that were permitted. A fight master needed to know the differences in procedure between one city and the next.

At the same time, these legal customs could change and evolve, albeit slowly. Regions like Swabia, Franconia and Bavaria were positioned in an important trade nexus at the headwaters of the Rhine and Danube rivers, and not far from the Alpine passes. When south German jurists looked beyond their borders for legal precedents, they drew inspiration not only from German-speaking lands to the north, but also from the whole Danube watershed and even from the Mediterranean world. Without a world-historical perspective, it is easy to mistake such objects as the faceted club or the tall spiked shield for unusual local innovations, when in fact they were variations of weapons that had existed elsewhere for a long time.

Finally, it is important to question whether all the plays portrayed in the *Fechtbücher* were contemporary with their authors. At times, the masters themselves seem to have been reconstructing their scenarios from historical sources, and perhaps even re-enacting them. Only by examining the precedents on a longer timeline can we determine which parts of their work were tradition, innovation, or re-enactment.

Bibliography

Manuscript sources

Eike von Repgow, Sachsenspiegel, Dresden, Sächsischen Landes- Staats- und Universitätsbibliothek, Mscr. Dresd. M.32.
———. Wolfenbüttel, Herzog August Bibliothek, Cod. Guelph. 3.1 Aug. 2.
Falkner, Peter, Kunste zu Ritterlicher Were, Vienna, Kunsthistorisches Museum, MS 5012.
Jean d'Ibelin, Assises de Jerusalem, Oxford, Bodleian Library MS Selden Supra 69.
Kal, Paulus, Fechtbuch, Munich, Bayerische Staatsbibliothek MS Cgm 1507.
New York, Pierpont Morgan Library MS M.457.
Paris, Bibliothèque nationale de France, MS fr. 13096. https://gallica.bnf.fr/ark:/12148/btv1b10533304x, accessed on 1 January 2019
Paris, Musée national du Moyen Âge, MS Cluny 23842.
René d'Anjou, Traité de la forme et devis comme on peut faire les tournois, Paris, Bibliothèque nationale de France, MS français 2695, https://gallica.bnf.fr/ark:/12148/btv1b84522067. image, accessed on 1 January 2019.
Schilling, Diebold, Spiezer Chronik, Bern, Burgerbibliothek, Mss. h. h. I. 16. https://www.e-codices.ch/en/list/one/bbb/ Mss-hh-I0016, accessed on 1 January 2019.

Talhoffer, Hans, Fechtbuch, Berlin, Stiftung Preußischer Kulturbesitz, MS 78.A.15.
———, Copenhagen, Kongelike Bibliotek, MS Thott 290.2.
———, Gotha, Universitäts- und Forschungsbibliothek, MS Chart.A.558.
———, Munich, Bayerische Staatsbibliothek, Codex iconographicus 394a.
Tours, Bibliothèque municipale, MS 0568.
Vatican, Biblioteca Apostolica Vaticana, MS Pal. lat. 1888. https://doi.org/10.11588/diglit.9740#0001, accessed on 1 January 2019.

Published primary sources

Agobard of Lyons, Agobardi Lvgdvnensis: Opera omnia, ed. L. Van Acker, Corpus Christianorum, Continuatio Medievalis LII (Turnhout: Brepols, 1981).
Antiquitates Italicae medii aevi, ed. Lodovico Antonio Muratori, 6 vols (Milan: Typographia Societatis Palatinae, 1738-1742).
Bovet, Honorat (aka Honoré Bonet), L'arbre de batailles de Honoré Bonet, ed. Ernest Nys (Brussels: C. Muquardt, 1883).
Capitularia Regum Francorum, ed. Alfred Boretius, Monumenta Germaniae Historica, Capitularia regum Francorum 1 (Hannover: Hahn, 1883).
Christine de Pizan, The Book of Deeds of Arms and Chivalry, ed. Charity Cannon Willard, trans. Sumner Willard (University Park, PA: Pennsylvania State University Press, 1999).
Chroniken der schwäbischen Städte: Augsburg, vol. 1 (Leipzig: S. Hirzel, 1875).
Crusius, Martin, Annales Suevici, 3 vols. (Frankfurt, Nikolaus Basse, 1595-6).
Deutschenspiegel und Augsburger Sachsenspiegel, ed. Karl August Eckhardt and Alfred Hübner, 2nd ed. Monumenta Germaniae Historica, Fontes iuris 3 (Hannover: Hahn, 1933).
Deutsche Stadtrechte des Mittelalters mit rechtsgeschichtliche Erläuterungen, ed. Ernst Theodor Gaupp (Breslau: Josef Max & Co., 1851).
Frederick II, Historia diplomatica Friderici Secundi, ed. Jean-Louis-Alphonse Bréholles & H. de Albertis de Luynes, 6 vols. (Paris: Plon, 1817-1871).
Fuero general de Navarra, ed. Pablo Ilarregui and D. Segundo Lapuerta (Pamplona: Imprenta Provincial, 1869).
Hessisches Urkundenbuch: Urkundenbuch zur Geschichte des Herrn von Hanau und der ehemaligen Provinz Hanau, ed. Heinrich Reimer, vol. 3: 1350-1375 (Leipzig: S. Hirzel, 1894).
Jung, Carl Ferdinand, ed. Miscellaneorum, vol. 1 (Frankfurt and Leipzig: n.p., 1739).

Knapp, Hermann, ed., Die Zenten des Hochstifts Würzburg: ein Beitrag zur Geschichte des süddeutschen Gerichtswesens und Strafrechts, mit Unterstützung der Savignystiftung, 2 vols. (Berlin: J. Guttentag, 1907).

Leges Langobardorum, ed. F. Bluhme & I.C. Bonensi, Monumenta Germaniae Historica, Leges 4 (Hannover: Hahn, 1868).

Lex Alamannorum, ed. Karl Lehman and Karl Augustus Eckhardt, 2nd ed., Monumenta Germaniae Historica, Leges nationum Germanicarum 5.1 (Hannover: Hahn, 1966).

Lex Ribvaria, ed. Franz Beyerle and Rudolf Buchner, Monumenta Germaniae Historica, Leges nationum Germicarum 3.2 (Hannover: Hahn, 1954).

Murḍa bin Alī al-Tarsūsī, "Tabṣirat Arbāb al-Albāb," in Claude Cahen, "Un Traité d'armurerie composé pour Saladin," Bulletin d'études orientales 12 (1947-8), 103-163.

Müller, Johann Joachim, ed. Des Heiligen Römischen Reichs Teutscher Nation Reichs Tags Theatrum, etc., 2 vols. (Jena: Johann David Werther, 1713).

Ordonnances des roys de France de la troisième race, ed. Eusèbe Jacques de Laurière, vol. 1 (Paris: Imprimerie Royale, 1723).

"Řád práva zemského—Ordo judicii terrae," in Codex juris Bohemici, ed. Hermengild Jireek, volume 2, part 2 (Prague: I.L. Kober, 1867), 198-255.

Rechtsbuch Kaiser Ludwigs des Bayern von 1346, ed. Wilhelm Volkert, Bayerische Rechtsquellen 24 (Munich: C.H. Beck, 2010).

Ruprecht von Freising, Freisinger Rechtsbuch, ed. Hans-Kurt Clauen (Weimar: Böhlau, 1941).

Sachsenspiegels erster Theil oder das Sächsiche Landrecht nach der Berliner Handschrift v. J. 1361, des, ed. C.G. Homeyer (Berlin: Ferdinand Dümmler, 1861).

Schwabenspiegel, oder schwäbisches Land- und Lehen-Rechtsbuch, ed. F.L.A. von Lassberg, (Tübingen: Ludwig Friedrich Fues, 1840).

Select Pleas of the Crown, ed. F.W. Maitland, Selden Society 1 (London: Bernard Quaritch, 1888).

Stadtbuch von Augsburg, inbesondere das Stadtrecht von 1276, ed. Christian Meyer (Augsburg: Butsch, 1872).

Stumpf, Johannes, Gemeiner loblicher Eydgnoschafft Stetten, Landen, Völckeren Chronick wirdinger Thaaten Beschreybung (Zurich: Christoffel Froschouer, 1548).

Teutsches Corpus Iuris, publici et privati, oder Corpus diplomaticus... ed. Joseph Stephan, vol. 1 (Ulm: Johann Conrad Bohler, 1717).

Weisthümer, ed. Jacob Grimm, 7 vols. (Göttingen: Dietrich, 1840-1878).

Wurzelmann, Mattheus, "Kampfgericht in Schwäbisch Hall," 1537, in Wurttembergische Jahrbücher: 1843, vol. 2, ed.

Johann Daniel Georg Memminger (Stuttgart: J.G. Cotta, 1846).

Zeerleder, Karl, ed. Urkunden für die Geschichte der Stadt Bern, 3 vols. (Bern: Stämpfli, 1853).

Zwickauer Rechtsbuch, ed. Günther Ullrich, (Weimar: H. Böhlau, 1941).

Zwierlein, Christian Jacob von, ed. Historisch-diplomatische Abhandlung von denen landesherrlichen Gerechtsamen des Hochfürstlichen Hauses Brandenburg über den Marktflecken Fürth, etc. (Anspach: Posch, 1771).

Secondary sources

Bartlett, Robert, Trial by Fire and Water: The Medieval Judicial Ordeal (Oxford: Clarendon Press, 1986).

Bergot, Louis-Patrick, "L'Apocalypse d'Isabelle de France (1313) et son lien avec un groupe de Bibles historiales," Questes: Revue pluridisciplinaire d'études médiévales 38 (2018): 63-79.

Boldog, Z., "A Prismatic Mace-Head from Dunaföldvár from the Time of the Mongol Invasion," Acta Archaeologica Academiae Scientiarum Hungaricae 63 (2012): 181-96.

Canel, A. "Le Combat judiciaire en Normandie" Mémoires de la Société des antiquaires de Normandie 22 (1836): 575-655.

Cavina, Marco. Il sangue dell'onore: Storia del duello (Rome: Laterza, 2005).

Denkstein, Vladimír, "Pavises of the Bohemian Type, II: The Origin and Development of Pavises in pre-Hussite Europe," in Sborník Národního Muzea v Praze, Series A-Historia 18:3-4 (1964): 149-194.

DeVries, Kelly, "The Introduction and Use of the Pavise in the Hundred Years War," Arms and Armour 4:2 (2007): 93-100.

Douet d'Arcq, Louis-Claude, Inventaire de sceaux, part 1, vol. 2 (Paris: H. Plon, 1867).

Elema, Ariella, "Trial by Battle in France and England," (doctoral dissertation University of Toronto, 2012).

Fărcaş, André-Octavian, "Maces in Medieval Transylvania between the Thirteenth and Sixteenth Centuries," (master's thesis, Central European University, 2016).

Fehr, Hans, Der Zweikampf (Berlin: Karl Curtius, 1908).

Fortner, Sandra, "'Kempflich angesprochen:' über Kampfgerichte und Kampfrecht," in André Schulze, ed., Mittelalterliche Kampfesweisen: Der Kriegshammer, Schild und Kolben (Mainz: Philipp von Zabern, 2007), pp. 10-22.

French, Thomas. York Minster: The St. William Window, Corpus Vitrearum Medii Aevi Great Britain Summary Catalogue 5 (Oxford: Oxford University Press, 1999).

Grimm, Jacob, Deutsche Rechtsalterthümer, 2nd ed. (Göttingen: Dietrich, 1854).

Hill, G. P., The Development of Arabic Numerals in Europe (Oxford: Clarendon Press, 1915).

Lea, Henry Charles, Superstition and Force. Book 2. 1866. Reprinted as The Duel and the Oath (Philadelphia: University of Pennsylvania Press, 1974).

Keen, Maurice, The Laws of War in the Late Middle Ages (London: Routledge & Keegan Paul, 2005).

Lewis, Suzanne, "The Apopcalypse of Isabella of France: Paris, Bibl. Nat. MS Fr. 13096," The Art Bulletin 72:2 (1990): 224-60.

Muhlberger, Steven, Deeds of Arms (Wheaton, IL: Freelance Academy Press, 2005).

Neumann, Sarah, Der gerichtliche Zweikampf: Gottesurteil, Wettstreit, Ehrensache, Mittelalter-Forschungen, Band 31 (Ostfildern, Germany: Jan Thorbecke, 2010).

Nottarp, Herman, Gottesurteilstudien (Munich: Kosel, 1956).

Pettengill, R.W., "The Source of an Episode in Heinrich von Neustadt's Apollonius," The Journal of English and German Philology 13:1 (Jan., 1914): 45-50.

Russell, M.J., "Accoutrements of Battle," Law Quarterly Review 99:3 (1983): 432-442.

Schultz, Alwin, Deutsches Leben im XIV und XV Jahrhundert (Vienna: Tempsky, 1892).

Sinclair, Niigan, "Appropriation or Appreciation," (lecture, OCAD University, 28 November, 2018).

Talaga, Maciej, "A Kampfschwert from the 15th Century—A Reinterpretation of the So-Called 'Teutonic Estoc' from the Princes Czartoryski Collection in Cracow, Poland," Acta Periodica Duellatorum 3 (2015): 7-27.

Unger, Friedrich Wilhelm, Der gerichtliche Zweikampf bei den germanischen Völkern (Göttingen: Vandenhoeck & Ruprecht, 1847).

White, Lynne, Jr., "Kyeser's 'Bellifortis': The First Technological Treatise of the Fifteenth Century," Technology and Culture 10:3 (1969): 436-41.

White, Stephen D., "Proposing the Ordeal and Avoiding It," in Cultures of Power: Lordship, Status and Progress in the Later Middle Ages, ed. Thomas Bisson (Philadelphia: University of Pennsylvania Press, 1995), 89-123.

Würdinger, Joseph, Beiträge zur Geschichte des Kampfrechts in Bayern (Munich: C. Wolf & Sohn, 1877).

6. Addenda and Esoterica in the Thott Talhoffer Codex

Christian Henry Tobler

The Thott Talhoffer manuscript is not alone in mixing fencing lore with other, sometimes seemingly-unrelated subjects.[454] Fencing treatises mix liberally with those on siege craft, veterinary science, astrology, magical formulae, name-based divination, and recipes for food, medicine, gunpowder, and the hardening of iron. In the case of the commonplace book Hs. 3227a, the fencing material is just one of these many subjects.

The 1459 Talhoffer manuscript is, however, first and foremost a *Fechtbuch*, so it raises the question: why is there other material included? The answer, in the most general sense, would lie with the holistic nature of medieval know-ledge. By why would Talhoffer assemble diverse topics together with his fighting know-ledge? If the book is a personal copy for Tal-hoffer's own use, it might be simply a way of compiling diverse things of interest into a single volume. More likely, in this author's opinion, is that the combination of the different subjects, coupled with the manuscript's high-quality artwork, argues in favor of the Thott codex being something produced for someone else, either on commission or as a form of advertisement of Talhoffer's wide-ranging knowledge to a prospective patron.

Barring further history of the manuscript surfacing, the above must remain speculation. What is clear is that the appearance of other subjects in conjunction with fencing lore in 15th century manuscripts is not particularly unusual. The famous 1405 *Bellifortis* war book (*Kriegsbuch*) of Conrad Kyeser (discussed elsewhere in this volume), and those deriving from it, finds itself copied into this and other *Fechtbücher*, such as Ludwig von Eyb's early 16th century work and the anonymous 15th century Codex Guelf 78.2 August 2°. The *Bellifortis* material is inserted after the recitation of Johannes Liechtenauer's *Zettel*, and before Talhoffer's *Ringen* material.

Other subject materials are appended to the Thott codex, and in singular fashion: the pages are inverted, requiring the reader to flip the book over to read them from folios now numbered 150ᵛ to 140ʳ. Whether this is the result of later rebinding or the original intention is unclear. We may be sure though that these folia are associated with the remainder of the manuscript, for Talhoffer's name appears on f. 148ᵛ, and the same scribe—Michel Rotwyler—is cited in both portions of the manuscript, along with the year 1459.

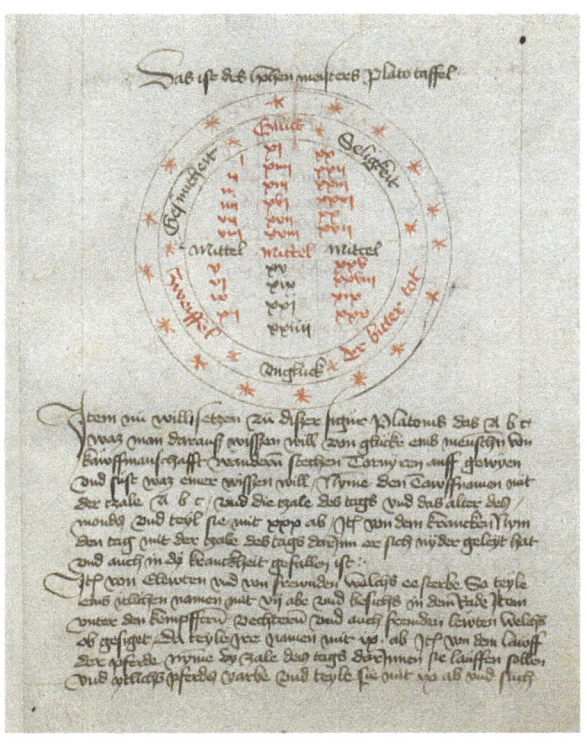

Figure 1: An onomantic diagram and associated text. Legend: Talhoffer, 1448. Gotha, Forschungsbibliothek Erfurt/Gotha, MS Chart.A.558, fol. 13ʳ.

---

[454] Before proceeding, I must acknowledge the translation work of JEFFREY HULL on this manuscript.

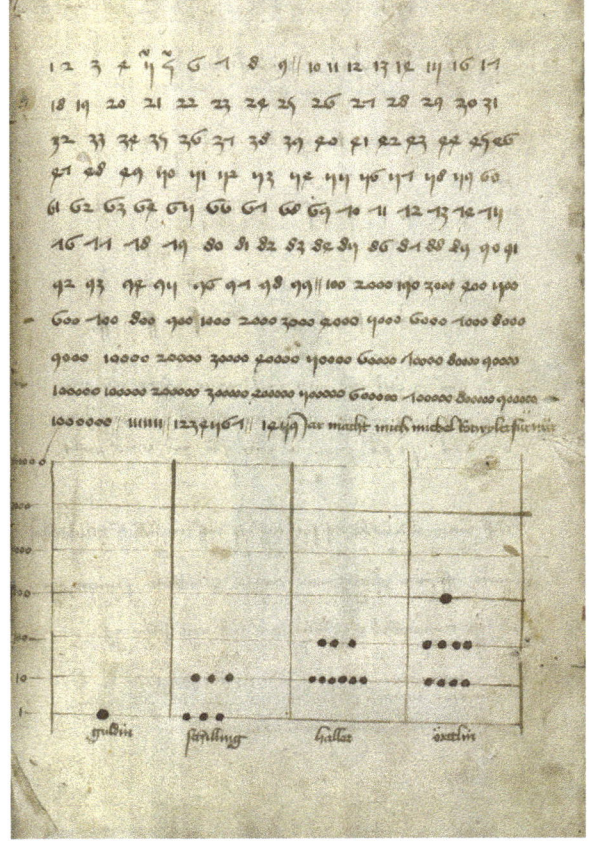

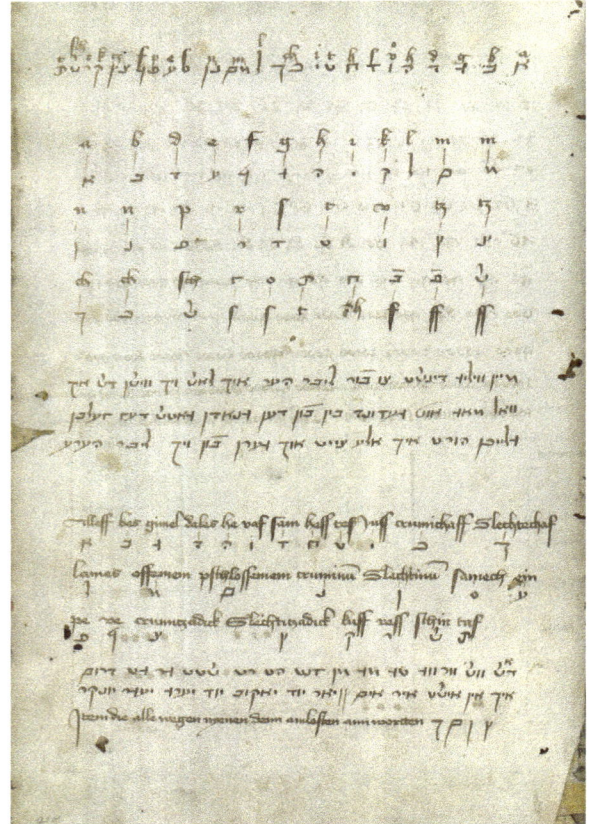

Four subjects are presented, arranged as either small treatises or tables. The first consists of tables describing the decimal system and currencies, followed by a tabular presentation of the Hebrew alphabet and renderings of some phrases and names employing it. And extensive treatise on planetary astrology follows. The final section is a treatise on physiology, partaking of much of the elemental lore that informs the astrological treatise.

Numeral and Currency Tables

The top half of folio 150ᵛ contains an exposition of the Indo-Arabic numeral system, detailing how base 10 numbers are written. In the first row of numbers, the numeral 5 is repeated, each with the Roman numeral 'V' atop it, probably to show variations in how it might appear when written; both variants appear in the numbers, first listing 1 to 100, that follow.

Following the last line of numbers appears the phrase 1459 *Jar macht mich michel rotwyler für wär* ("In the year 1459, Michel Rotwyler truly made me"), indicating the likely scribe of the Thott codex.

The lower half of 150ᵛ displays a table, its vertical axis marked in powers of 10, of currency. The table describes how one guldin (likely the late medieval south German *gulden*) equates to 33 schillings, 360 haller (named for Schwäbisch Hall), or 1440 "öxttlin". Dieter Bachmann notes that the correspondence of the guildin to 33 schillings is erroneous and that this should be "30 (depending on the currency even 40 or 24, but not 33) Schilling".[455] I am no expert in this area, and therefore must defer to him in this matter, but I can imagine there might have been local variations in currency that are not well documented today.

Hebrew Alphabet

Folios 150ʳ-149ᵛ are a short lesson on the Hebrew alphabet, written right to left across the top of 150ʳ, with the corresponding German

---

[455] BACHMANN, DIETER. Comments on www.wiktenauer.com and www.schwertfechten.de

letters (or combination thereof) atop each letter. In some cases, more than one Hebrew character accords with the same German sound, dependent on its place within a word. Likely because of this, a more spread out tabular form follows, written left to right, breaking out these variants.

Some difficult, and perhaps sample, German text, rendered in Hebrew characters, follows, interrupted by another table describing how to pronounce the names of the letters of the Hebrew alphabet; it appears to be about loyal service to one's lord as well as leading a pious life. The page concludes with the curious statement "Item, they always naysay then at last reply", followed by the name, in Hebrew characters "Ch[e]mn[i]tz". Chemnitz is a city in Saxony, but may here be a name—perhaps the name of an author?

The facing page, 149ᵛ, renders three names in both German and Hebrew characters: Hans Talhoffer, Claus Pflieg[er], and Michel Rotwyler, the manuscript's scribe. Claus Pflieger may be an artist, or perhaps another scribe working with Rotwyler.

This folio is dominated by a figure, identified as a Jew by his conical *Judenhut* hat and the yellow *rota* annular badge. The figure points with its right hand to the facing page, with the rubricated text above him stating "Here the Jew teaches Hebrew." Two Jews are named among the previous page's text—Jacob, and Jerg Juncker—but it is unclear if this figure is meant to be one of them or simply a generic Jewish character.

Whether this lesson is intended to familiarize the reader with Hebrew or employ its alphabet for the purpose of encryption, is uncertain.

Planetary Astrology

A simple system of astrology came into vogue in the early 15th century, perhaps even as early as the late 14th century. A form of 'popular astrology', it is distinct from the mathematically complex attempts at divination via the heavens found in more sophisticated astrological treatises of the time. This was the lore of the *Planetenkinder*, or "Children of the Planets", and it described human behavior and tendencies based on what planet's influence a person was born under.

The presentation of the *Planetenkinder* differs from source to source, and they are found in diverse works: dedicated astrolog-

Figure 2: A Jewish teacher. Legend: Talhoffer, 1459. Copenhagen, Det Kongelige Bibliotek, Thott 290 2°, fol. 149ᵛ.

Figure 4 (op. top): Numbering system. Legend: Talhoffer, 1459. Copenhagen, Det Kongelige Bibliotek, Thott 290 2°, fol. 150ᵛ.

Figure 3 (op. bottom): Hebrew alphabet. Legend: Talhoffer, 1459. Copenhagen, Det Kongelige Bibliotek, Thott 290 2°, fol. 150ʳ.

Figure 5: Saturn and Jupiter, as depicted in Bellifortis.
Legend: Bellifortis, *Konrad Kyeser, 1405. Göttingen, Niedersächsische Staats- und Universitätsbibliothek Göttingen, Cod. Philos. 63, fol. 6ʳ, 7ʳ.*

ical texts, almanacs, commonplace books, and, notably, as a part of the aforementioned *Bellifortis*. Sometimes this takes the form of illustration, accompanied by bits of Latin and/or German doggerel; at other times, more extensive commentary appears.

Although much of *Bellifortis* is included in the Thott codex, the planetary treatise here is not the simple one found in that war book. *Bellifortis' Planetenkinder* consists of anthropomorphized illustrations of the planets, along with some simple Latin verses. In this copy of Talhoffer, no illustrations are provided, but considerable German text is devoted to each planet, plus some additional remarks on the motion of the heavens.

The cosmogony of late Antiquity through the Renaissance recognized seven planets, which included the sun and moon. In this pre-Copernican system, the planets were held to move within concentric celestial spheres which, from the earth outward, were ordered, according how long they took to traverse the sky: Luna (the moon), Mercury, Venus, Sol (the sun), Mars, Jupiter, and finally Saturn, beyond which the sphere of the unmoving stars lies. Beyond the planetary and stellar spheres was the celestial realm of Heaven.

Each planet was associated with signs of the zodiac (one each for the sun and moon, two each for the remaining five) and described in terms of their elemental characteristics and the qualities they bestowed upon their "children"—those born

under their influence. This treatise, like others on the *Planeten-kinder*, begins with Saturn, the outermost planet, and works its way inward through the other six planets.

While the treatise acknowledges Saturn as the uppermost of the planets, and the "highest god" of the Romans, it decries his unvirtuous and miserable nature, assigning the elementary qualities of cold and dryness (those belonging to Earth, the lowest of the four classical elements). His zodiacal signs are Capricorn and Aquarius, and he is the laziest of planets, taking 30 years to complete his course in the sky. Saturn's children are said to be sickly, sad, and prone to robbery and vulgarity.

Jupiter is described in more positive terms. Warm and moist in nature, like the element of Air, he rules the signs of Sagittarius and Pisces. "Lucky and virtuous", his children are moderate, merry, and honorable.

Hateful and warlike Mars partakes of the element Fire's hot and dry nature. Ruling over Aries and Scorpio, his children are wrathful and quarrelsome. Another manuscript featuring the *Planetenkinder*, M III 36[456] in the Universitätsbibliothek Salzburg, agrees with this assessment, providing this blunt condemnation of Mars' nature:

Figure 6: Mars and the Sun, as depicted in Bellifortis. *Legend:* Bellifortis, *Konrad Kyeser, 1405. Göttingen, Niedersächsische Staats- und Universitätsbibliothek Göttingen, Cod. Philos. 63, fol. 8ʳ, 9aʳ.*

---

[456] This work includes the *Planetenkinder*, but also the Seven Liberal Arts and a medieval cosmogony diagram.

Figure 7: Venus and Mercury, as depicted in Bellifortis. Legend: Bellifortis, Konrad Kyeser, 1405. Göttingen, Niedersächsische Staats- und Universitätsbibliothek Göttingen, Cod. Philos. 63, fol. 9bʳ, 10ʳ.

"*Mars, much malice he commands,*
in his houses thus he stands
His children angry courage bear:
the blood of monks they do not spare."

The fourth planet down, Sol receives the greatest praise in astrological lore. While sharing in Mars' fiery nature, he is lusty, beautiful, merry, and shines upon all the other planets. Like the moon, the sun rules but one astrological house, Leo, and grants wisdom, artistry, and health to his children, his face "illuminating Mankind".

Cold and wet, and therefore watery in nature, Venus is unchaste and hasty, and yet a lucky planet, a counterbalance to Mars' wicked ways. The text notes this planet's dual roles as morning and evening star, calling it Lucifer at dawn and Vespera at dusk. Commanding Taurus and Libra, her children are beautiful, drawn to music and dance.

This treatise calls Mercury fiery, but then goes on to describe how its nature is varied and tempered by its proximity in the sky to other—"good and bad"—planets, hence, mercurial. Mercury rules the signs Virgo and Gemini, and imparts the qualities of speed, charm, and wit to his children.

The last planet, and the hastiest in its travel through the heavens, is Luna—the moon. Ruling over Cancer, the moon is counted among the watery planets, but also changeable, like its waxing and waning face. Her children are moody, unstable, and have a tendency toward laziness.

"It is to be known of the seven planets and their natures that God has ordained them and that he is above the stars." So begins a discussion, following the exposition on the seven planets, of how the planets' natures and their elemental qualities (hot, cold, wet, and dry) are related to the signs they rule. This in turn affects men's natures, making some men hot and dry (fiery) and therefore "good-natured and bold", while others, of cold and wet nature, are talkative and lazy. The planetary treatise then concludes with a discussion of the sun's course in the sky, detailing its time in the zodiacal houses in the twelve months of the year.

Anatomy

The last section of this inverted portion of the Thott Talhoffer codex, a short anatomical treatise, begins with the rubricated title "Here begins a book on how the body is internally arranged". The text describes the organs, based on the day's understanding.

The brain's function of control of the limbs and interaction with the five senses is not too distantly removed from our modern understanding. However, when the text describes the heart, we find a mixture of physical reality and ancient (mis)understanding—the heart circulating blood and warmth, but also containing the soul. The treatise goes on to say that the liver is responsible for the body's hydration, drawing liquid from the stomach and, in another example of early anatomical misconceptions, that the kidneys are the source of semen in the male.

Next, the anonymous author describes the function of the epiglottis in ensuring food passes to the stomach rather than the lungs. The treatise also affirms that the lungs take in cold air and expels hot air, which could internally smother a man. Some sense of how disease is spread is noted, both from others already sickened, but also from "bad air". The description of the organs concludes with the stomach, wherein, as the author avers, the food is prepared for the supply of the other organs, akin to a cooking pot.

The anatomical discourse concludes with references to another work by a "Master Allmonser", "*Panthagin*", which discusses the organs of the body in the same elemental quality vein seen in the astronomical treatise. Cold organs, which include

Figure 8: The Moon, as depicted in Bellifortis.
Legend: Bellifortis, *Konrad Kyeser, 1405. Göttingen, Niedersächsische Staats- und Universitätsbibliothek Göttingen, Cod. Philos. 63, fol. 10ᵛ.*

185

the bones, bladder, stomach, and bowels, and those the author affirms have no blood in them. Hot organs are those with blood: the heart, liver, spleen, and flesh. This distinction then informs a short discussion on the flow of bodily fluids and the processing of waste in the intestines.

Again, I must draw upon comments by Dieter Bachmann, who notes that "Master Allmonsor" is likely Manṣūr ibn Ilyās (hence, "Al-Mansur"), who was a 14th century Persian anatomical authority, but that the *Panthagin* (or, more correctly the *Liber Pantegni*) is not his work, but a Hellenistic treatise. Bachmann further affirms that the lore in this manuscript is from Antiquity, and not from Al-Mansur.[457]

---

[457] BACHMANN, ibid.

About the Contributors (in alphabetical order)

Paul Becker is a German historian, fencing teacher, officer and coach, who since his youth has devoted himself to the training and research of historical fencing arts and their history. From 2005-2018 Paul Becker was an officer in the German Army in the rank of a captain. There he studied history with a focus on the Middle Ages to a Master of Arts. In the military he was also a teacher and instructor in the fields of combat, fighting, sport, shooting, communication and security policy. Since the end of 2018, he has been running his fencing school IN MOTU full-time, developing and selling products for historical fencing. He also works as a freelance historian, where he was responsible for the exhibition "Iron Bestselling Sword and Armour in Craft and Society" as curator. His current research focuses on German-speaking fencing masters of the 14th-15th century and the fencing didactics of the German fencing art of this period.

Michael Chidester is the Editor-in-Chief of Wiktenauer, a component of the non-profit HEMA Alliance. Michael is a Research Scholar of the Meyer Freifechter Guild, a founding member of the Society for Historical European Martial Arts Studies (SHEMAS), a member of the Western Martial Arts Coalition (WMAC), and a Lifetime Member of the HEMA Alliance. He has lectured on historical martial arts across North America and Europe and authored several books, including *The Illustrated Meyer* and *The Long Sword Gloss of GNM Manuscript 3227a*.

Ariella Elema is an archivist and information sciences professional. She received a PhD in medieval studies from the University of Toronto in 2012, and an MLIS from Western University in 2018. Her PhD dissertation, *Trial by Battle in France and England*, won the Canadian Society of Medievalists' 2013 Leonard Boyle Dissertation Prize. Ironically, she practices *armizare*.

Rebecca L.R. Garber received her PhD in German Languages and Literatures from the University of Michigan in 1999, with a specialization in medieval studies. She taught for 3 years at Wayne State University and then at a private high school before turning to translation as a full-time occupation. After joining a stage combat troupe in Michigan in 2006, she began translating combat manuals with CHEMAS (the Cambridge Historical European Martial Arts Study group), where her expertise in Medieval German and Latin was more useful than in most aspects of modern life. This is her second published HEMA translation, the first being "Florius de Arte Luctandi" in *The Flower of Battle of Master Fiore Furlano de'I Liberi* (with Kendra Brown).

Dierk Hagedorn works for the IT subsidiary of a large German airline for numerous creative issues such as user experience, interface and corporate design. He is an instructor for the long sword at Hammaborg, a club for historical fencing in Hamburg; he is also a passionate wearer of armour. He has transcribed 30 of the approximately 80 surviving fight books in the German language and made them available on the website of his club or published them in book form. Among his numerous publications are editions about the manuscripts by Peter von Danzig, Lew the Jew, Hans Talhoffer and Jörg Wilhalm.

Daniel Jaquet is a medievalist, with a background in literary studies and interest in history of science and material culture in the early modern period. He received his PhD in history at the University of Geneva in 2013. He taught at the University of Geneva and Lausanne (2008-2015). He was a visiting scholar at Max Planck Institute for History of Science (Berlin, 2015-2016), and an associate researcher at the Renaissance Centre of the University of Tours (2016-2017). His teaching and research specialisations are history of warfare, duelling, martial practices and knowledge transmission in pragmatical literature at the end of the Middle Age and the beginning of the Renaissance. He is currently the coordinator of the research project "Martial Culture in Medieval Towns" (Swiss National Science Foundation, 2018-2022) and the head of research and public engagement at the Military Museum of the Castle of Morges.

Christian Henry Tobler is Principal Instructor for the Selohaar Fechtschule, a school for historical fencing, and is the author of several books on German medieval martial arts, including *Fighting with the German Longsword* and *In Service of the Duke*. He has taught at numerous events in the US and abroad and is a founder of the Chivalric Martial Arts Association and a Companion of the Seven Swords. He lives in Connecticut with his wife, along with a considerable collection of reproduction arms and armour. He has worked for many years, in various capacities, in the analytical instrumentation industry.

The Personal Manuscript of Hans Talhoffer

You have the companion volume, but what about the facsimile? Talhoffer was the first facsimile produced by HEMA Bookshelf, and copies are still available.

This facsimile is printed on cotton rag paper to simulate the texture and handling of Medieval paper and is sewn following the current quire structure. Bound in tan pigskin with 9 mm boards and blind tooling based on the manuscript, this is an exacting replica that will make you feel like you're holding Talhoffer's original.

Your collection won't be complete without this extraordinary work, so order today!

hemabookshelf.com/talhoffer-facsimile

www.ingramcontent.com/pod-product-compliance
Lightning Source LLC
Chambersburg PA
CBHW041647120626
46551CB00017B/2340